This book provides scientists and engineers with a comprehensive understanding of composite materials, which form an important class of engineering materials. In describing their production, properties and usage, the book crosses the borders of many disciplines, from the physics and chemistry of the materials themselves, to their design and applications in engineering.

This new edition has been greatly revised and expanded, and analyses both the theoretical and practical aspects of the subject. Topics covered include: the properties of fibres, matrices, laminates and interfaces; elastic deformation, stress and strain, strength, fatigue crack propagation and creep resistance; deterioration under environmental conditions. Coverage has been enlarged to include polymeric, metallic and ceramic matrices and reinforcement in the form of long fibres, short fibres and particles. New chapters describe toughness and thermal properties, the fabrication and applications of composites, and an outline of the use of the Eshelby method.

Designed primarily as a teaching text, this book has evolved from courses given by the authors to final year undergraduate students of materials science and materials engineering. It will also be of interest to undergraduates and postgraduates in chemistry, physics, and mechanical engineering. In addition it will be an excellent source book for those involved in academic and technological research on materials.

AN INTRODUCTION TO COMPOSITE
MATERIALS, Second Edition

Cambridge Solid State Science Series

EDITORS

Professor D. R. Clarke
Department of Materials Science and Engineering,
University of California, Santa Barbara

Professor S. Suresh
Richard E. Simmons Professor, Department of Materials Science and Engineering,
Massachusetts Institute of Technology

Professor I. M. Ward FRS
IRC in Polymer Science and Technology,
University of Leeds

AN INTRODUCTION TO COMPOSITE MATERIALS

Second Edition

D. HULL

Emeritus Professor
University of Cambridge

AND

T. W. CLYNE

Reader in Mechanics of Materials,
Department of Materials Science and Metallurgy,
University of Cambridge

CAMBRIDGE
UNIVERSITY PRESS

Published by the Press Syndicate of the University of Cambridge
The Pitt Building, Trumpington Street, Cambridge CB2 1RP
40 West 20th Street, New York, NY 10011-4211, USA
10 Stamford Road, Oakleigh, Melbourne 3166, Australia

First published 1981

Second edition 1996

A catalogue record for this book is available from the British Library

Library of Congress cataloguing in publication data

Hull, Derek.
An introduction to composite materials / D. Hull and T. W. Clyne. – 2nd ed.
p.   cm. – (Cambridge solid state science series)
Includes bibliographical references.
ISBN 0-521-38190-8 (hardcover). – ISBN 0-521-38855-4 (pbk.)
1. Composite materials. I. Clyne, T. W. II. Title. III. Series.
TA418.9.C6H85 1996
620.1'18–dc20   96–5701 CIP

ISBN 0 521 38190 8 hardback
ISBN 0 521 38855 4 paperback

Transferred to digital printing 2003

KW

Contents

Preface

From the preface to First Edition

A book on composite materials which is fully comprehensive would embrace large sections of materials science, metallurgy, polymer technology, fracture mechanics, applied mechanics, anisotropic elasticity theory, process engineering and materials engineering. It would have to cover almost all classes of structural materials from naturally occurring solids such as bone and wood to a wide range of new sophisticated engineering materials including metals, ceramics and polymers. Some attempts have been made to provide such an over-view of the subject and there is no doubt that the interaction between different disciplines and different approaches offers a fruitful means of improving our understanding of composite materials and developing new composite systems.

This book takes a rather narrower view of the subject since its main objective is to provide for students and researchers, scientists and engineers alike, a physical understanding of the properties of composite materials as a basis for the improvement of the properties, manufacturing processes and design of products made from these materials. This understanding has evolved from many disciplines and, with certain limitations, is common to all composite materials. Although the emphasis in the book is on the properties of the composite materials as a whole, a knowledge is required of the properties of the individual components: the fibre, the matrix and the interface between the fibre and the matrix.

The essence of composite materials technology is the ability to put strong stiff fibres in the right place, in the right orientation with the right volume fraction. Implicit in this approach is the concept that in making the composite material one is also making the final product. This means that there must be very close collaboration between those who design composite materials at the microscale and those who have to design and manufacture the final engineering component.

Composite materials can be studied at a number of different levels each of which requires a different kind of expertise. The method of approach depends on the objectives of the investigation. Thus, the development of a composite material to resist a corrosive environment, while maintaining its physical and mechanical properties, is primarily an exercise in selecting fibres, resins and interfaces which resist this environment and is within the expertise of chemists, physicists and materials scientists. In contrast, the engineer who has to design a rigid structure, such as an aerodynamic control surface on an aircraft or a pressure pipeline, is more concerned with the macroscopic elastic properties of the material. He uses anisotropic elasticity theory and finite element analysis to design an optimum weight or optimum cost structure with the desired performance characteristics. The disciplines in these two examples barely overlap and yet it is important for the physical scientist to understand the nature of the design problem and for the engineer to appreciate the subtleties of the material he uses in design. This book goes some way towards building the bridge between these widely different approaches and should be of value to all scientists and engineers concerned with composite materials. Naturally, each group will look to other texts for an in-depth treatment of specific aspects of the subject.

Preface to Second Edition

In the 15 years since the first edition was published, the subject of composite materials has become broader and of greater technological importance. In particular, composites based on metallic and ceramic matrices have received widespread attention, while the development of improved polymer-based systems has continued. There have also been significant advances in the understanding of how composite materials behave. Furthermore, the wider range of composite types has led to greater interest in certain properties, such as those at elevated temperature. We therefore decided to produce a major revision of the book, covering a wider range of topics and presenting appreciably deeper treatments in many areas. However, because the first edition has continued to prove useful and relevant, we have retained much of its philosophy and objectives and some of its structure. Throughout the book, emphasis is given to the principles governing the behaviour of composite materials. While these principles are applicable to all types of composite material, examples are given illustrating how the detailed characteristics of polymeric-, metallic- and ceramic-based systems are likely to differ.

The first chapter gives a brief overview of the nature and usage of composite materials. This is followed by two chapters covering, firstly, the types of reinforcement and matrix materials and, secondly, geometrical aspects of how these two constituents fit together. The next three chapters are concerned with the elastic deformation of composites. Chapter 4 deals with material containing unidirectionally aligned continuous fibres, loaded parallel or transverse to the fibre axis. This is extended in Chapter 5 to laminates made up of bonded stacks of thin sheets, each having the fibres aligned in a particular direction. The following chapter covers discontinuously reinforced composites, containing short fibres or particles. Equations are presented in these chapters which allow prediction of elastic properties, but the emphasis is on pictorial representation of the concepts involved and it is not necessary to follow the mathematical details in order to understand and use the results.

Chapter 7 is concerned with the interface between matrix and reinforcement. This covers the nature of the interfacial bond in various systems and the measurement and control of bond strength. The interface often has an important influence on properties related to inelastic deformation and failure of composites. Treatment of this aspect is divided between the next two chapters, the first dealing with stress levels at which various deformation and damage processes occur and the second concerning energy absorption and quantification of the toughness of composite materials. The thermal behaviour of composites is described in Chapter 10, which includes thermal stresses, creep and thermal conduction.

The last two chapters are largely independent of the rest of the book. The first of these gives a brief survey of the manufacturing methods used to produce components from various types of composite. This aspect is particularly important, since the material and the component are commonly made in the same operation, at least for long-fibre composites. This calls for close integration between the processes of material specification and component design. This requirement is also highlighted in the final chapter, covering applications. The intention here is to identify some of the advantages and problems of using composites, by means of a series of illustrative case histories, rather than to give a systematic survey. To aid in the use of the book, a nomenclature listing is given as an appendix.

The contents have largely evolved from undergraduate courses we have given and, as with the first edition, the book is intended as a teaching aid at this level. It should also prove useful for scientists and engineers working with composite materials and for those engaged in research in this area. At the end of each chapter, a list of references is given, many of them relevant to specific points made in the text. These references should serve as useful sources of further detailed information at the research level. They need not, in general, be consulted by undergraduates studying the subject for the first time. A further point concerning additional sources relates to computer-assisted learning. Software packages are now available which allow both interactive exploration of elementary topics and calculation of composite properties not easily obtained from analytical equations. In many cases, these can serve as both teaching and research tools. One such package, entitled 'Mechanics of Composite Materials' (Clyne & Tanovic, published by the Institute of Materials in 1995 and by Chapman and Hall, as part of the MATTER software series, in 1996), is largely based on material in this book.

We would like to acknowledge the support of many colleagues in Cambridge and Liverpool Universities. Collaboration with and suggestions from W. J. Clegg, A. Kelly and P. J. Withers have been particularly useful. Stimulation and support from past and present students in our research groups, particularly in the Materials Science Department at Cambridge, have also been very helpful. In addition, we are indebted to all those who have provided us with micrographs and unpublished information. These are acknowledged in the text and figure captions. We would also like to acknowledge the financial and moral support we have received for our own research work on composites, in particular from the Engineering and Physical Sciences Research Council, Alcan International, British Petroleum, Ford Motor Company, Imperial Chemical Industries, National Physical Laboratory, Pechiney, DRA Farnborough, Pilkington, Rolls Royce, T & N, and Scott–Bader. We have had extensive scientific contact with various people from these and other organisations, which has been of considerable benefit to us. We are also grateful to Brian Watts, our copy editor, for his painstaking work and many useful suggestions, and to the editorial staff at CUP for their cooperation and efficiency in producing this book.

Finally, we would like to thank our wives, Pauline and Gail, for their invaluable support during the preparation of this book.

D. Hull
T. W. Clyne
1996

1

General introduction

Composites make up a very broad and important class of engineering materials. World annual production is over 10 million tonnes and the market has in recent years been growing at 5–10% per annum. Composites are used in a wide variety of applications. Furthermore, there is considerable scope for tailoring their structure to suit the service conditions. This concept is well illustrated by biological materials such as wood, bone, teeth and hide; these are all composites with complex internal structures designed to give mechanical properties well suited to the performance requirements. Adaptation of manufactured composite structures for different engineering purposes requires input from several branches of science. In this introductory chapter, an overview is given of the types of composite that have been developed.

1.1 Types of composite material

Many materials are effectively composites. This is particularly true of natural biological materials, which are often made up of at least two constituents. In many cases, a strong and stiff component is present, often in elongated form, embedded in a softer constituent forming the **matrix**. For example, wood is made up of fibrous chains of cellulose molecules in a matrix of lignin, while bone and teeth are both essentially composed of hard inorganic crystals (hydroxyapatite or osteones) in a matrix of a tough organic constituent called collagen (Currey 1983). Commonly, such composite materials show marked **anisotropy** – that is to say, their properties vary significantly when measured in different directions. This usually arises because the harder constituent is in fibrous form, with the fibre axes preferentially aligned in particular directions. In addition, one or more of the constituents may exhibit inherent anisotropy

1

as a result of their crystal structure. In natural materials, such anisotropy of mechanical properties is often exploited within the structure. For example, wood is much stronger in the direction of the fibre tracheids, which are usually aligned parallel to the axis of the trunk or branch, than it is in the transverse directions. High strength is required in the axial direction, since a branch becomes loaded like a cantilevered beam by its own weight and the trunk is stressed in a similar way by the action of the wind. Such beam bending causes high stresses along its length, but not through the thickness.

In making artificial composite materials, this potential for controlled anisotropy offers considerable scope for integration between the processes of material specification and component design. This is an important point about use of composites, since it represents a departure from conventional engineering practice. An engineer designing a component commonly takes material properties to be isotropic. This is often inaccurate even for conventional materials; for example, metal sheet usually has different properties in the plane of the sheet from those in the through-thickness direction, as a result of crystallographic texture (preferred orientation) produced during rolling – although such variations are in many cases relatively small. In a composite material, on the other hand, large anisotropies in stiffness and strength are possible and must be taken into account during design. Not only must variations in strength with direction be considered, but the effect of any anisotropy in stiffness on the stresses set up in the component under a given external load should also be taken into account. The material should be produced bearing in mind the way it will be loaded when it is made into a component. Thus, the processes of material production and component manufacture must be integrated into a single operation. This, of course, is exactly what happens when biological materials are produced.

There are several different types of composite. Examples of typical microstructures for the three main classes, grouped according to the nature of the matrix, are shown in Fig. 1.1. Most composites in industrial use are based on polymeric matrices; thermosets and thermoplastics. These are usually reinforced with aligned ceramic fibres, such as glass or carbon. They commonly exhibit marked anisotropy, since the matrix is much weaker and less stiff than the fibres. More recently, there has been considerable interest in metal matrix composites (MMCs), such as aluminium reinforced with ceramic particles or short fibres, and titanium containing long, large-diameter fibres. The property enhancements being sought by the introduction of reinforcement are often less pronounced

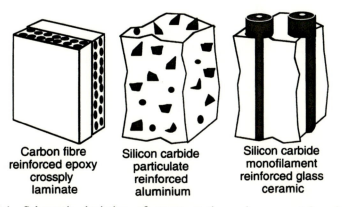

| Carbon fibre reinforced epoxy crossply laminate | Silicon carbide particulate reinforced aluminium | Silicon carbide monofilament reinforced glass ceramic |

Fig. 1.1 Schematic depiction of representative polymer, metal and ceramic matrix composites.

than for polymers, with improvements in high-temperature performance or tribological properties often of interest. While various industrial applications have been developed or are being explored for MMCs, their commercial usage is still quite limited when compared to that of polymer composites (PMCs). Finally, composites based on ceramic materials (CMCs) are also being studied. The objective here is usually to impart toughness to the matrix by the introduction of other constituents, since the stiffness and strength are unlikely to be much affected. Such materials are still, for the most part, in the early stages of development, partly because they are rather difficult to manufacture.

In considering the formulation of a composite material for a particular type of application, it is important to consider the properties exhibited by the potential constituents. The properties of particular interest are the stiffness (Young's modulus), strength and toughness. Density is of great significance in many situations, since the mass of the component may be of critical importance. Thermal properties, such as expansivity and conductivity, must also be taken into account. In particular, because composite materials are subject to temperature changes (during manufacture and/or in service), a mismatch between the thermal expansivities of the constituents leads to internal residual stresses. These can have a strong effect on the mechanical behaviour. Some representative property data are shown in Table 1.1 for various types of matrix and reinforcement, as well as for some typical engineering materials and a few representative composites. Inspection of these data shows that some attractive property combinations (for example, high stiffness/strength and low density) can

Table 1.1 *Overview of properties exhibited by different classes of material*

Type of material (example)	Density ρ (Mg m^{-3})	Young's modulus E (GPa)	Tensile strength σ (MPa)	Fracture toughness K_c (MPa $\sqrt{m}$)	Thermal conductivity K (W m^{-1} K^{-1})	Thermal expansivity α (10^{-6} K^{-1})
Thermosetting resin (epoxy)	1.25	3.5	50	0.5	0.3	60
Engineering thermoplastic (nylon)	1.1	2.5	80	4	0.2	80
Rubber (polyurethane)	1.2	0.01	20	0.1	0.2	200
Metal (mild steel)	7.8	208	400	140	60	17
Construction ceramic (concrete)	2.4	40	20	0.2	2	12
Engineering ceramic (alumina)	3.9	380	500	4	25	8
Wood (load // grain) (spruce) (load ⊥ grain)	0.6 / 0.6	16 / 1	80 / 2	6 / 0.5	0.5 / 0.3	3 / 10
General PMC (in-plane) (chopped strand mat)	1.8	20	300	40	8	20
Adv. PMC (load // fibres) (APC-2) (load ⊥ fibres)	1.6 / 1.6	200 / 3	1500 / 50	40 / 5	200 / 40	0 / 30
MMC (Al–20%SiC$_p$)	2.8	90	500	15	140	18

be obtained with composites. An outline of how such properties can be predicted from those of the individual constituents forms an important part of the contents of this book.

1.2 Design of composite materials

Choosing the composition and structure of a composite material for a particular application is not a simple matter. The introduction of reinforcement into a matrix alters all the properties. It is also necessary to take account of possible changes in the microstructure of the matrix resulting from the presence of the reinforcement. The generation of residual stresses from differential thermal contraction during manufacture may also be significant. Before considering such secondary effects, it is useful to take a broad view of the property combinations obtainable from different composite systems. This can be visualised using *property maps*. An example is presented in Fig. 1.2. This shows a plot of Young's modulus, E, against density, ρ. A particular material (or type of material) is associated with a point or a region. This is a convenient method of comparing the property combinations offered by potential matrices and reinforcements with those of alternative conventional materials.

Attractive matrix/reinforcement combinations can be identified by deriving a *'merit index'* for the performance required, in the form of a specified combination of properties. Appropriate models can then be used to place upper and lower bounds on the composite properties involved in the merit index, for a given volume fraction of reinforcement. The framework for such predictions has been set out by Ashby (Ashby 1993). An example is shown in Fig. 1.3 for three different fibres and a polymer matrix. The shaded areas joining the points corresponding to a fibre to that of the matrix represent the possible combinations of E and ρ obtainable from a composite of the two constituents concerned. (The density of a composite is given simply by the weighted mean of the constituents; the stiffness, however, can only be identified as lying between upper and lower bounds – see Chapter 4 – unless more information is given about fibre orientation.)

Also shown on Fig. 1.3 are lines corresponding to constant values of the ratios E/ρ, E/ρ^2 and E/ρ^3. These ratios represent the merit indices to be maximised to obtain minimum component weight consistent with a maximum permissible deflection for different component shapes and loading configurations. For example, the lightest square-section beam able to support a given load without exceeding a specified deflection is

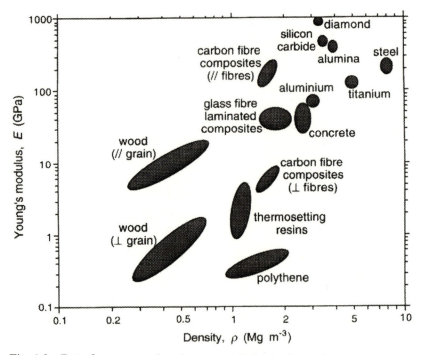

Fig. 1.2   Data for some engineering materials in the form of areas on a map of
Young's modulus E against density ρ.

the one made of the material with the largest value of E/ρ^2. It can be seen
from the figure that, while the introduction of carbon and silicon carbide
fibres would improve the E/ρ ratio in similar fashions, carbon fibres
would be much the more effective of the two if the ratio E/ρ^3 were the
appropriate merit index.

1.3 The concept of load transfer

Central to an understanding of the mechanical behaviour of a composite
is the concept of load sharing between the matrix and the reinforcing
phase. The stress may vary sharply from point to point (particularly with
short fibres or particles as reinforcement), but the proportion of the
external load borne by each of the individual constituents can be gauged
by volume-averaging the load within them. Of course, at equilibrium, the
external load must equal the sum of the volume-averaged loads borne by

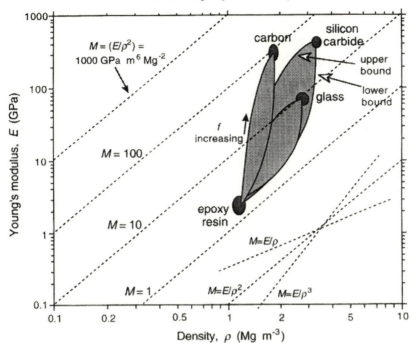

Fig. 1.3 Predicted map of Young's modulus E against density ρ for composites of glass, carbon or silicon carbide fibres in a matrix of epoxy resin. The shaded areas are bounded by the axial and transverse values of E predicted for the composite systems. The diagonal dotted lines represent constant values of three merit indices (E/ρ, E/ρ^2 and E/ρ^3). For the E/ρ^2 case, several lines are shown corresponding to different values of the ratio.

the constituents[†] (e.g. the matrix and the fibre). This gives rise to the condition

$$f\bar{\sigma}_{\mathrm{m}} + (1-f)\bar{\sigma}_{\mathrm{f}} = \sigma_{\mathrm{A}} \qquad (1.1)$$

governing the volume-averaged matrix and fibre stresses ($\bar{\sigma}_{\mathrm{m}}$, $\bar{\sigma}_{\mathrm{f}}$) in a composite under an external applied stress σ_{A}, containing a volume fraction f of reinforcement. Thus, for a simple two-constituent composite under a given applied load, a certain proportion of that load will be carried by the fibre and the remainder by the matrix. Provided the response of the composite remains elastic, this proportion will be inde-

[†]In the absence of an externally applied load, the individual constituents may still be stressed (due to the presence of residual stresses), but these must balance one another according to Eqn (1.1).

pendent of the applied load and it represents an important characteristic of the material. It depends on the volume fraction, shape and orientation of the reinforcement and on the elastic properties of both constituents. The reinforcement may be regarded as acting efficiently if it carries a relatively high proportion of the externally applied load. This can result in higher strength, as well as greater stiffness, because the reinforcement is usually stronger, as well as stiffer, than the matrix. Analysis of the load sharing which occurs in a composite is central to an understanding of the mechanical behaviour of composite materials.

References and further reading

Ashby, M. F. (1993) Criteria for selecting the components of composites, *Acta Metall. Mater.*, **41**, 1313–35

Chou, T. W. (1993) *Microstructural Design of Fiber Composites*. Cambridge University Press: Cambridge

Currey, J. D. (1983) Biological composites, in *Handbook of Composites, vol. 4 – Fabrication of Composites*. A. Kelly and S. T. Mileiko (eds.), Elsevier: New York, pp. 501–64.

Kelly, A. (ed.) (1994) *Concise Encyclopaedia of Composite Materials*. Pergamon: Oxford

Weeton, J. W., Peters, D. M. and Thomas, K. L. (eds.) (1987) *Engineers Guide to Composite Materials*. ASM: Metals Park, Ohio

2
Fibres and matrices

In this chapter, the underlying science of fibres and matrices is described. Some specific examples are given to illustrate the key factors involved. A wide range of reinforcements, mostly in the form of fibres, is now available commercially. Their properties can be related directly to the atomic arrangement and the defect content of the reinforcement, which must be controlled in the manufacturing processes. Matrices may be based on polymers, metals or ceramics. The choice of matrix is related to the required properties, the intended applications of the composite and the method of manufacture. The properties of the matrix depend on microstructure which, in turn, depends on manufacturing route and subsequent thermal and mechanical treatments. Certain properties of the composite may be sensitive to the nature of the reinforcement/matrix interface; this topic is covered in detail in Chapter 7.

2.1 Reinforcements

Many reinforcements are now available, some designed for a particular matrix system. A selection is listed in Table 2.1. Typical properties of fibres are given in Table 2.2. All have high stiffness and relatively low density. Carbon, glass and aramid fibres are now used extensively in polymer matrix composites. Carbon fibres are also important for carbon/carbon composites. Ceramic fibres, whiskers and particles can be used to reinforce metal and ceramic matrices.

2.1.1 Carbon fibres

In a graphite single crystal, the carbon atoms are arranged in hexagonal arrays, stacked in a regular ABABAB... sequence. The atoms in these

Fibres and matrices

Table 2.1 *Some common types of reinforcement*

Form	Size (μm)		Fabrication route	Examples
	d	*L*		
Monofilaments (large-diameter single fibres)	100–150	∞	CVD onto core fibres (e.g. of C or W)	SiC (SCS-6[TM]) Boron
Multifilaments (tows or woven rovings with up to 14 000 fibres per strand)	7–30	∞	Precursor stretching; pyrolysing; melt spinning	Carbon (HS & HM) Glass Nicalon[TM] Kevlar[TM] 49 FP[TM] alumina
Short fibres (staple fibres aggregated into blankets, tapes, wool, etc.)	1–10	50–5000	Spinning of slurries or solutions, heat treatment	Saffil[TM] Kaowool Fiberfrax[TM]
Whiskers (fine single crystals in loose aggregates)	0.1–1	5–100	Vapour phase growth/reaction	SiC Al_2O_3
Particulate (loose powder)	5–20	5–20	Steelmaking byproduct; refined ore; sol-gel processing, etc.	SiC Al_2O_3 B_4C TiB_2

basal planes are held together by strong covalent bonds, with only weak van der Waals forces between them. The basic crystal units are therefore highly anisotropic; the in-plane Young's modulus (normal to the *c*-axis) is about 1000 GPa, while that perpendicular to the basal planes (parallel to the *c*-axis) is only 35 GPa.

Carbon fibres, which are typically about 8 μm in diameter, consist of small crystallites of 'turbostratic' graphite, one of the allotropic forms of carbon. Turbostratic graphite closely resembles graphite single crystals, except that the layer planes are not regularly packed in the *c*-axis direction. The structure is represented schematically in Fig. 2.1. To obtain high axial modulus and strength, good alignment of the basal planes parallel to the fibre axis is required. The arrangement of the layer planes in the cross-section of the fibre is also important, since it affects the transverse and shear properties. Compilations are available (Weeton *et al.* 1987) of the properties exhibited by commercially used carbon fibres.

Table 2.2 *Fibre properties*

Fibre	Density ρ (Mg m^{-3})	Young's modulus E (GPa)	Poisson's ratio ν	Tensile strength σ_* (GPa)	Failure strain ϵ_* (%)	Thermal expansivity α (10^{-6} K^{-1})	Thermal conductivity K (W m^{-1} K^{-1})
SiC monofilament	3.0	400	0.20	2.4	0.6	4.0	10
Boron monofilament	2.6	400	0.20	4.0	1.0	5.0	38
HM[a] carbon	1.95	axial 380 radial 12	0.20	2.4	0.6	axial −0.7 radial 10	axial 105
HS[b] carbon	1.75	axial 230 radial 20	0.20	3.4	1.1	axial −0.4 radial 10	axial 24
E-glass	2.56	76	0.22	2.0	2.6	4.9	13
Nicalon[TM]	2.6	190	0.20	2.0	1.0	6.5	10
Kevlar[TM] 49	1.45	axial 130 radial 10	0.35	3.0	2.3	axial −6 radial 54	axial 0.04
FP[TM] fibre	3.9	380	0.26	2.0	0.5	8.5	8
Saffil[TM]	3.4	300	0.26	2.0	0.7	7.0	5
SiC whisker	3.2	450	0.17	5.5	1.2	4.0	100
Cellulose (flax)	1.0	80	0.3	2.0	3.0	—	—

[a] High modulus
[b] High strength

Fibre
axis

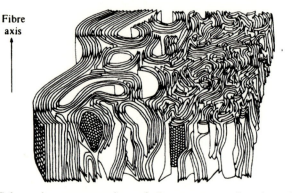

Fig. 2.1   Schematic representation of the structure of carbon fibres. (From Bennett and Johnson, 1978).

There are three main routes for producing carbon fibres.

(a) *From polyacrylonitrile fibres.* This process was developed at Rolls Royce and the Royal Aircraft Establishment, Farnborough, UK, with important contributions by Watt and co-workers (1966–67). It is now the preferred route for producing most high-modulus carbon fibres. The starting material is the polymer polyacrylonitrile (PAN). This closely resembles polyethylene, but with one of the two hydrogen atoms on every other carbon backbone atom replaced by a nitrile ($-C\equiv N$) group. Bulk PAN is drawn down to a fibre and stretched to produce alignment of the molecular chains. When the stretched fibre is heated, the active nitrile groups react to produce a **ladder polymer**, consisting of a row of six-membered rings – see Fig. 2.2. While the fibre is still under tension, it is heated in an oxygen-containing environment. This causes further chemical reaction and the formation of cross-links between the ladder molecules.

The oxidised PAN is then reduced to give the carbon ring structure, which is converted to turbostratic graphite by heating at higher temperatures. The fibres usually have a thin skin of circumferential basal planes and a core with randomly oriented crystallites. The effect of final heat-treatment temperature on the properties of PAN-based carbon fibres is illustrated by the data in Fig. 2.3. By suitable choice of the final temperature, some control is possible over the Young's modulus and the tensile strength. Fibres can be produced with Young's modulus values referred to as standard (230–240 GPa), intermediate (250–300 GPa) and high (350–500 GPa). In general, high-modulus fibres have lower strains to

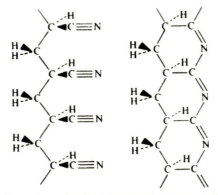

Fig. 2.2   Transformation of a PAN molecule into a rigid ladder polymer.

failure than low-modulus fibres – see Table 2.2. The transverse modulus, perpendicular to the fibre axis, is usually about 3–10% of the axial modulus. Other properties, such as thermal expansion coefficient and thermal conductivity, also tend to be quite highly anisotropic.

(b) *From mesophase pitch*. This process was originally reported by Otani (1965) and has since been developed by several workers. Pitch is a complex mixture of thousands of different species of hydrocarbon and heterocyclic molecules. If it is heated above 350 °C, condensation reactions occur, leading to large, flat molecules which tend to align parallel to one another. This is often termed 'mesophase pitch', a viscous liquid exhibiting local molecular alignment (i.e. a liquid crystal). This liquid is then rapidly extruded through a multi-hole spinneret to produce 'green' yarn. During this process, hydrodynamic effects in the orifice generate overall alignment of the molecules. The yarn is made infusible by oxidation at temperatures below its softening point. These fibres can be converted thermally, without any applied tension, into a graphitic fibre possessing a high degree of axial preferred orientation. The basal planes are usually oriented radially, as well as being aligned along the fibre axis. This conversion is carried out at about 2000 °C. The resulting structures are highly graphitic – more so than for fibres produced from polymeric fibres. This has certain implications for the thermophysical properties. For example, pitch-based fibres can exhibit very high thermal conductivities of $\sim 1000\,\mathrm{W\,m^{-1}\,K^{-1}}$ (see Kowalski 1987). Such values, which are much higher than for PAN-based fibres, are above those typical of copper. This is an important advantage for certain applications, such as with the carbon/carbon composites used for aircraft brakes (§12.9).

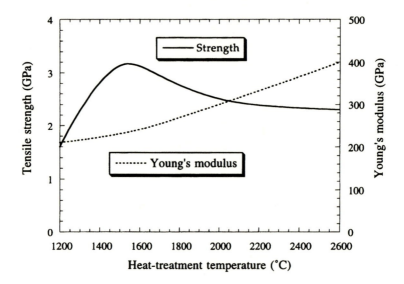

Fig. 2.3 Effect of heat-treatment temperature on the strength and Young's modulus of carbon fibres produced from a PAN precursor. (From Moreton *et al.* 1967).

(c) *By pyrolytic deposition.* Carbon fibres can be produced by pyrolytic deposition of hydrocarbons from the gas phase. Methane, benzene and naphthalene have been used, with deposition temperatures of about 1100 °C. The process was originally described by Oberlin *et al.* (1976). A thin tube of carbon is first formed on a suitable substrate and this then grows by a surface diffusion mechanism. The turbostratic carbon is formed in concentric stacks parallel to the fibre axis. Technical details have been given by Tibbetts *et al.* (1987). Fibre lengths of up to 50 mm have been reported. In these short lengths, and with poor control over fibre diameter, there has been little commercial use of these fibres, but they can be highly graphitised and have properties which might be attractive for some applications.

2.1.2 Glass fibres

Most glass fibres are based on silica (SiO_2), with additions of oxides of calcium, boron, sodium, iron and aluminium. These glasses are usually amorphous, although some crystallisation may occur after prolonged heating at high temperatures, leading to a reduction in strength. Typical compositions of three types of glass popular for composites are

Table 2.3 *Glass fibre compositions and properties*

	E-glass	C-glass	S-glass
Composition (%)			
SiO_2	52.4	64.4	64.4
$Al_2O_3 + Fe_2O_3$	14.4	4.1	25.0
CaO	17.2	13.4	—
MgO	4.6	3.3	10.3
$Na_2O + K_2O$	0.8	9.6	0.3
B_2O_3	10.6	4.7	—
BaO	—	0.9	—
Properties			
ρ (Mg m^{-3})	2.60	2.49	2.48
K (W m^{-1} K^{-1})	13	13	13
α (10^{-6} K^{-1})	4.9	7.2	5.6
σ_* (GPa)	3.45	3.30	4.60
E (GPa)	76.0	69.0	85.5
T_{max}(°C)	550	600	650

given in Table 2.3. The most commonly used, E-glass (E for electrical), draws well and has good strength, stiffness, electrical and weathering properties. In some cases, C-glass (C for corrosion) is preferred, having better resistance to corrosion than E-glass, but a lower strength. Finally, S-glass (S for strength) is more expensive than E-glass, but has a higher strength, Young's modulus and temperature resistance.

Glass fibres are produced by melting the raw materials in a reservoir and feeding into a series of platinum bushings, each of which has several hundred holes in its base. The glass flows under gravity and fine filaments are drawn mechanically downwards as the glass extrudes from the holes. The fibres are wound onto a drum at speeds of several thousand metres per minute. Control of the fibre diameter is achieved by adjusting the head of the glass in the tank, the viscosity of the glass (dependent on composition and temperature), the diameter of the holes and the winding speed. The diameter of E-glass is usually between 8 and 15 μm.

The strength and modulus are determined primarily by the atomic structure. Silica-based glasses consist primarily of covalently bonded tetrahedra, with silicon at the centre and oxygen at the corners. The oxygen atoms are shared between tetrahedra, leading to a rigid three-dimensional network. The presence of elements with low valency, such as Ca, Na and K, tends to break up this network by forming ionic bonds with oxygen atoms, which then no longer bond the tetrahedra together. Addition of such elements therefore tends to lower the stiffness and strength, but improves the formability. In contrast to carbon fibres, the properties of glass fibres are isotropic. Thus, the axial and transverse Young's moduli are the same. The strength (see Table 2.2) depends on processing conditions and test environment. Freshly drawn E-glass fibres, provided they are handled very carefully to avoid surface damage, have a strength of 3.5 GPa and the variability in strength is small (see §2.2.2). The strength falls in humid air, owing to the adsorption of water on the surface. A sharper decrease occurs if the surface is exposed to mineral acids.

A major factor determining the strength is the *damage* which fibres sustain when they rub against each other during processing operations. To minimise this damage, glass fibres are usually treated with a *size* at an early stage in manufacture. This is a thin coating applied to the fibres by spraying with water containing an emulsified polymer. The size serves several purposes: (a) to protect the surface of the fibres from damage, (b) to bind the fibres together for ease of processing, (c) to lubricate the fibres so that they can withstand abrasion during subsequent processing operations, (d) to impart anti-static properties and (e) to provide a chemical link between the glass surface and the matrix to increase the interface bond strength. These are complex requirements and the size normally contains the following constituents: (i) a film-forming polymer, such as polyvinyl acetate, to provide protection during handling, (ii) a lubricant, usually a small molecule of some type, and (iii) a coupling agent to provide a bond between fibre and matrix, often an organosilane (see §7.3.1).

2.1.3 Organic fibres

The most important high modulus polymer fibres have been developed from aromatic polyamides and are called *aramid* fibres. They were first developed by Du Pont with the trade name '*Kevlar*^{TM}'. The principles involved are best understood by considering fibres based on a simple polymer, *polyethylene*. Chain-extended polyethylene single crystals consist of straight zig-zag carbon backbone chains, fully aligned and closely packed. These have a Young's modulus of about 220 GPa parallel to the chain axis. Fairly good chain alignment can be achieved in a fibre by drawing and stretching and a modulus of 70 GPa has been achieved. As for carbon fibres, the modulus normal to the fibre axis is much less than that along the fibre axis.

Unlike polyethylene, which crystallises largely by chain folding, aramid fibres are derived from polymer molecules with a high degree of aromaticity (containing benzene rings), which exhibit liquid crystalline behaviour in solution. The molecules act as rigid rods, which readily align parallel to each other to form ordered domains. When solutions of these molecules are subjected to shear, the ordered domains tend to orient in the direction of flow. Aramid fibres are produced by extrusion and spinning processes. A solution of the polymer in a suitable solvent (e.g. sulphuric acid) is passed through a spinneret to develop a high degree of orientation. After removal of residual solvent, further alignment and ordering of the molecules is achieved by a thermal annealing treatment. The polymer molecules form rigid planar sheets, as illustrated in Fig. 2.4 for poly(*p*-phenylene terephthalamide) molecules. The chain-extended molecules are aligned parallel to the axis of the fibre and there is weak inter-chain hydrogen bonding between the molecules. Aramid fibres are highly anisotropic (see Table 2.2) and, because of the weak inter-chain bonding, they readily split into much finer fibrils and microfibrils. This is commonly seen during damage to composites containing Kevlar™ fibres.

There is a wide range of natural organic fibres which are potentially useful as reinforcement. The most common natural fibre is cellulose, which is formed by polymerisation of glucose molecules. The most common natural composite is timber, which is essentially composed of crystalline cellulose fibres in a matrix of amorphous or partially crystalline hemi-cellulose and lignin. These are arranged in various complex configurations in the cell walls of different plants. Cellulose fibres (microfibrils) can be extracted from a range of plants, such as cotton, flax and jute, as well as timber. It can be seen from the data for flax fibres shown in Table 2.2 that the properties of cellulose fibres compare fairly well with those of many artificial fibres, particularly if density is important. Although production of composites based on such fibres is currently very limited, it may increase in the future.

2.1.4 Silicon carbide

Silicon carbide has a structure similar to diamond, and is attractive as a reinforcement in view of its low density and high stiffness and strength, combined with good thermal stability and thermal conductivity. It is much easier to synthesise than diamond and can be produced readily in large quantities in a crude form such as powder. Production of fibres

Fibres and matrices

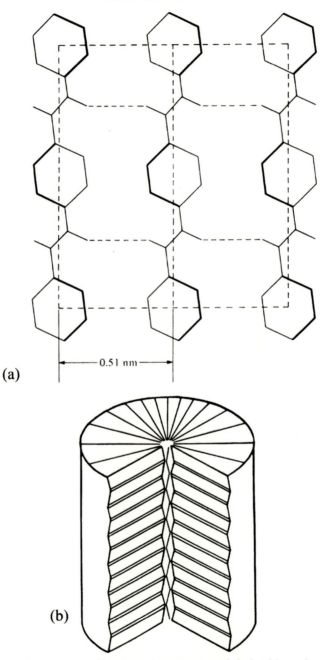

(a)

(b)

Fig. 2.4   (a) Planar array of poly-*p*-phenylene terephthalamide molecules show-ing inter-chain hydrogen bonding. (b) Supramolecular structure of Kevlar[TM] 49, depicting the radially arranged pleated system. (From Dobb 1980).

in large quantities is more problematic, and several different routes have been developed.

(i) *CVD monofilaments*. Large-diameter (~ 100–$150\,\mu m$) fibres, often termed monofilaments, are made by chemical vapour deposition (CVD) onto fine core fibre substrates, usually carbon ($\sim 30\,\mu m$ diameter) or tungsten ($\sim 10\,\mu m$ diameter). The core fibre is fed continuously through a reaction chamber. A gaseous carbon-containing silane, such as methyltrichlorosilane (CH_3SiCl_3), is passed through the chamber. The core fibre is heated, usually by passing an electrical current through it, and the gas dissociates thermally at the fibre surface to deposit the SiC. Surface layers, designed to improve the resistance to handling damage and the compatibility with the matrix (usually a metal, intermetallic or ceramic), are often deposited in a second reactor. For example, graphitic layers are commonly applied. There has been extensive study of the nature and effects of interfacial reactions with metallic matrices, particularly titanium alloys (Martineau *et al.* 1984). The process is employed commercially for production of boron fibres, as well as SiC (Lepetitcorps *et al.* 1988).

(ii) *PCS multifilaments*. Fibres which are primarily SiC are made by a polymer precursor route analogous to the PAN-based method for carbon fibres (§2.1.1). Fibres about $15\,\mu m$ in diameter are produced in a similar manner, using ***polycarbosilane*** (PCS) as a precursor. The best known fibre produced by this route, '*Nicalon*[TM]', was first developed by Yajima in Japan (1976–78). Polycarbosilane is produced in a series of chemical steps involving the reaction of dichlorodimethylsilane with sodium to produce polydimethylsilane, which yields polycarbosilane on heating in an autoclave. This is spun into fibres, which are then pyrolised at temperatures up to 1300 °C. The final fibre contains a substantial proportion of SiO_2 and free carbon as well as SiC. Similar processes are used for the production of silicon nitride (Si_3N_4) fibres (Okamura *et al.* 1987).

(iii) *Whiskers*. The best mechanical properties are expected from single-crystal reinforcement in the form of fibrous whiskers. This is confirmed by the data in Table 2.2. Whiskers are small (~ 0.1–$1\,\mu m$ diameter), virtually defect-free, single-crystal rods. They have been produced in small quantities for research purposes in a wide variety of materials. The growth mechanisms involved have been known for many years. Commercial production of large quantities has been limited to SiC whiskers, mostly produced in Japan. For example, α-SiC whiskers have been grown from vaporised SiC using a catalyst and β-SiC whiskers have been deposited by hydrogen reduction of CH_3SiCl_3. Processing details, and the effects of various characteristics on properties, are given by Levitt (1970). The very high strength of whiskers, which are fully crystalline, is a direct result of the absence of dislocations and surface flaws. The TEM micrograph shown in Fig. 2.5 illustrates that fine-scale twins are often present in these whiskers. The use of whiskers, which was actively developed for reinforcement of ceramics and metals, has declined somewhat as a consequence of concern (Birchall *et al.* 1988) about the hazards of handling a potentially carcinogenic material (with a similar size range to asbestos fibres).

(iv) *Particulate.* Equiaxed powder particles offer advantages in terms of cost and ease of handling and processing. Such material is used in a wide variety of composite materials, often simply as a cheap filler. For example, many engineering polymers and rubbers contain additions of

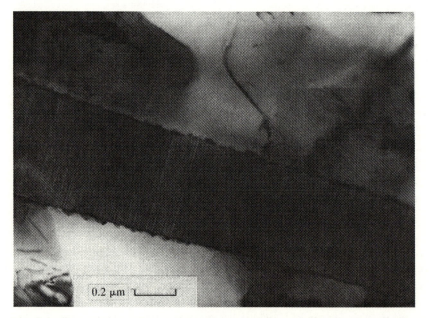

Fig. 2.5   Transmission electron (TEM) micrograph of a SiC whisker, showing the fine internal twin structure.

powders such as talc, clays, mica, silica and silicates. In many cases, the mechanical properties of the powder particles are of little concern and a cheap material is used. In other situations, density reduction might be a primary aim, as with the additions of hollow glass or ceramic microspheres. However there are cases, notably for MMCs, where particles of high stiffness and strength are needed, often having good chemical compatibility with the matrix. The most common example of this is SiC grit, already in widespread commercial use as an abrasive, for reinforcement of aluminium alloys (McDanels 1985). A major concern here is with cost, since interest is centred on high-volume components for which economic factors are a major consideration (see Chapter 12).

2.1.5 Alumina and aluminosilicates

Inorganic oxide fibres are, for the most part, made up of combinations of alumina and silica. Fibres containing approximately 50 wt% of each of these (aluminosilicates) comprise by far the greatest tonnage of refractory fibres and are in extensive use for high-temperature insulation purposes. Such fibres are glassy. Alumina fibres with much lower silica contents, which are crystalline, are more expensive to manufacture, but have greater resistance to high temperature and higher stiffness and strength.

(i) *Multifilaments.* Continuous alumina fibres are made by a slurry process. An aqueous suspension of alumina particulate is extruded into fibre form and then fired. Such a fibre, made up of fine grains (0.5 μm) of α-alumina and having a diameter of 20 μm, is manufactured by Du Pont under the trade name of '*FP*^{TM} *fibre*'. It has been quite extensively used for metal matrix composites (Dhingra 1980a,b). Economic factors, processing limitations and, for titanium-based composites, problems of interfacial reaction, have militated against extensive commercial use.

(ii) *Short fibres.* Aluminosilicate fibres are usually manufactured from kaolin or a related clay material, which is melted and either poured into jets of gas or fed onto a rapidly rotating disk. Both of these processes produce fibrous material, with some particulate ('shot'), the spinning process tending to favour longer fibres. For production of alumina fibres, the low viscosity and high melting point of alumina preclude simple melt spinning. However, short fibres can be made by the spinning of viscous, concentrated solutions of aluminium compounds which are precursors to the oxide (Birchall 1983). Such fibres are manufactured under the trade name of '*Saffil*™' ('safe filament'), having been marketed initially as a replacement for asbestos. Typically, about 5% of silica is included, in order to stabilise the δ crystal structure of alumina. This silica is concentrated at the grain boundaries and free surface of the fibre, favouring a very localised surface attack in some metallic matrices. The fibre has found extensive use as a reinforcement in aluminium-based composites (Clyne *et al.* 1985, Abis 1989).

2.2 The strength of reinforcements

Information on the tensile strengths of fibres is given in Tables 2.3 and 2.3. A number of factors should be borne in mind when quoting tensile strength figures; some of these are outlined briefly below.

2.2.1 Thermal stability

In the absence of air or other oxidising environments, carbon fibres possess exceptionally good high-temperature properties. In Fig. 2.3 it was shown that fibres are heated at temperatures up to 2600 °C. Carbon fibres retain their superior properties to well above 2000 °C. For applications involving polymer matrices, this property cannot be used because most matrices cannot be used above about 200 °C (see §2.3.1). Carbon fibres have not been extensively employed for metal matrix composites, since they tend to react chemically during fabrication. However, in carbon/carbon composites, the high-temperature capability is fully exploited, providing the composite is isolated from oxidising environments. In the presence of oxygen, carbon fibres rapidly degenerate at elevated temperatures.

Bulk silica-based glasses have softening temperatures around 850 °C, but the strength and modulus of E-glass decreases rapidly above 250 °C. This is not usually a problem for polymer composites, but it eliminates glass fibres from use in many inorganic composites. Although the thermal stability of aramid fibres is inferior to both glass and carbon, it is again adequate in most polymer matrix systems. Apart from retaining the properties during service at elevated temperatures, it is essential that deterioration in properties does not occur during manufacturing operations. The properties of glass are reversible with temperature, but aramid fibres suffer irreversible deterioration owing to changes in internal structure. Care must be taken that the heating during curing of thermosetting resins and in melt processing of thermoplastics does not degrade the properties of fibres such as Kevlar™.

Oxide, nitride and carbide ceramics have very high melting points; for example, Al_2O_3 melts at about 2050 °C and SiC sublimes at about 2700 °C. Such fibres retain their properties to high temperatures, providing there are no adverse reactions with the matrix material, either during processing or in service. Much current research is directed towards the need to protect fibres under these conditions and a variety of coating treatments have been developed (see §7.3).

In addition to temperature and oxidation effects, aramid fibres are prone to photo-degradation on exposure to sunlight. Both visible and ultraviolet light have an effect and lead to discoloration and reduction of mechanical properties. The degradation can be avoided by coating the surface of the composite material with a light-absorbing layer.

2.2.2 Compressive strength

No reference is made in Table 2.2 to the axial compressive strength of fibres. This is difficult to measure and is best inferred from the behaviour of composites fabricated with the fibres. The axial compressive strength of unidirectional composites is controlled by the buckling modes of the fibres. This is in contrast to the tensile strength, which is dependent primarily on the tensile strength of the fibres. Under compressive loads, failure may occur in several different modes. These include crushing or shearing, but one of the most likely modes in a composite is by **Euler buckling** of individual fibres. Such buckling occurs when a rod under compression becomes unstable against lateral movement of the central region. The applied stress for the onset of buckling, in the case of a cylindrical rod, is given by

$$\sigma_{*b} = \frac{\pi^2 E}{16}\left(\frac{d}{L}\right)^2 \tag{2.1}$$

where E is the Young's modulus, d is the diameter and L is the length of the rod. Buckling will be favoured when the rod has a high **aspect ratio** (L/d). The buckling of fibres in a composite is more difficult to predict, since some lateral stabilisation is provided by the matrix, even when it has a relatively low stiffness. (In the absence of this, as for example with **ropes**, an assembly of fibres has very little resistance to buckling.) In any event, an important point to note is that, for compressive loading of a composite specimen, large-diameter fibres will tend to exhibit greater resistance to local buckling and this will often increase the resistance to macroscopic failure.

However, fibre diameter is not the only factor to consider. For example, the axial compressive strength of a unidirectional lamina made from Kevlar™ fibres is only 20% of its tensile strength, whereas for carbon and glass (which have similar diameters), the strengths are approximately equal in tension and compression. This arises because of the structure of the aramid fibre (see Fig. 2.4). In compression, elastic deformation (of the strong covalent bonds) only occurs to a limited degree before the weak van der Waals forces between adjacent molecules is overcome and local fibrillation and damage occur. This leads to buckling and kink formation at low applied loads, as shown by the micrograph in Fig. 2.6.

2.2.3 Fibre fracture and flexibility

In tension, most fibres tend to fracture in a brittle manner, without any yield or flow. Carbon, glass and ceramic fibres are almost completely brittle and fracture without any reduction in cross-sectional area. This

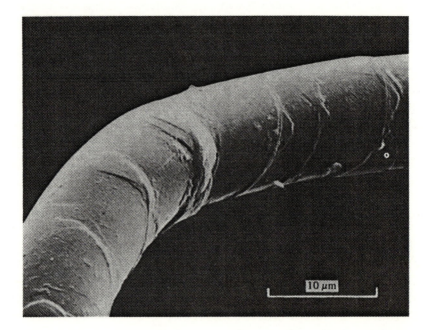

Fig. 2.6 Scanning electron (SEM) micrograph of a Kevlar™ 49 (aramid) fibre, showing deformation bands on the compression side of a sharp bend.

can be seen in Fig. 2.7(a)–(c). In contrast, aramid fibres fracture in a ductile manner, although the overall strain to failure is still small. Pronounced necking precedes fracture and final separation occurs after a large amount of local drawing – see Fig. 2.7(d). Fracture usually involves fibrillation of the fibres.

The diameter of the fibres has a large effect on the ease with which they can be deformed and bent. This is important in operations where fibres are fed through eyes and over bobbins, as in weaving, knitting and filament winding. The same is true of moulding and mixing operations where

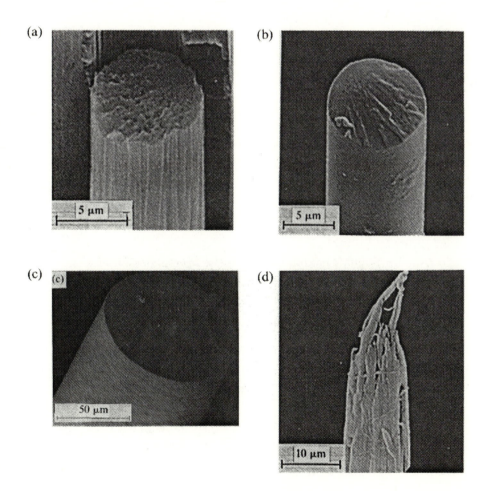

Fig. 2.7 SEM micrographs of fibres fractured in tension: (a) carbon, (b) E-glass, (c) silicon carbide monofilament, and (d) aramid.

Fibres and matrices

Table 2.4 *Fibre flexibility data*

Fibre	Diameter d (μm)	Young's modulus E (GPa)	Flexibility κ/M (GPa^{-1} mm^{-4})	Fracture strength σ_* (GPa)	Maximum curvature κ_{max} (mm^{-1})
SiC monofilament	150	400	1×10^2	2.4	0.08
Nicalon^{TM}	15	190	2.1×10^6	2.0	1.4
Kevlar^{TM} 49	12	130	7.6×10^6	3.0	3.8†
E-glass	11	76	1.8×10^7	2.0	4.8
HM carbon	8	390	1.3×10^7	2.2	1.4
HS carbon	8	250	2.0×10^7	2.7	2.7
Saffil^{TM}	3	300	8.4×10^8	2.5	5.5
SiC whisker	1	450	4.5×10^{10}	5.0	22.2

†Aramid fibres tend to become damaged easily in compression; at a considerably smaller curvature than this, they will have become permanently deformed by kink-band formation – see Fig. 2.6

fibres are intimately mixed and subsequently extruded or injection moulded using screw-fed machines. The *flexibility* of a fibre can be expressed in terms of the moment, M, required to bend a fibre with a circular cross-section to a given curvature, κ, (the reciprocal of the radius of curvature)

$$M = \frac{\pi E \kappa d^4}{64} \tag{2.2}$$

where E is the Young's modulus and d is the fibre diameter. The flexibility, defined as κ/M, is dominated by d but also depends on E. A comparison of the flexibilities of carbon, glass, aramid and ceramic fibres is given in Table 2.4.

Bending of fibres results in high surface tensile stresses, which can lead to fibre fracture. Assuming elastic deformation, the surface tensile stress is given by

$$\sigma = \frac{E \kappa d}{2} \tag{2.3}$$

For a fibre with a given fracture strength, σ_*, there will be a maximum curvature which the fibres can sustain before fracture occurs, given by

$$\kappa_{max} = \frac{2\sigma_*}{Ed} \tag{2.4}$$

Values of κ_{max} are given in Table 2.4, neglecting any statistical variation in fracture strength along the fibre (see §2.2.4). Glass fibres are much more tolerant of bending than carbon or ceramic fibres. Large-diameter monofilaments are very difficult to weave or form in any way, since they tend to break at a radius of curvature around 10 mm. Staple fibres and whiskers, on the other hand, are very unlikely to be broken by normal handling operations. A further point should be noted with respect to aramid fibres. As mentioned above, they tend to have a low compressive strength. Bending produces high surface compressive stresses as well as tensile stresses. Long before the bending curvature is sufficient to cause tensile failure, the compressive region of the fibre undergoes yielding by the development of deformation (or kink) bands. This results in a permanent deformation, as illustrated in Fig. 2.6.

2.2.4 A statistical treatment of fibre strength

Most fibres are essentially brittle. That is to say, they sustain little or no plastic deformation or damage up to the point when they fail catastrophically. Such materials do not, in general, have a well-defined tensile strength. The stress at which they fail usually depends on the presence of flaws, which may occur randomly along the length of a fibre. In fact, the high strength of fibres, compared to corresponding bulk materials, is often attributable to the absence of large flaws. Nevertheless, a population of flaws is expected along the length of fibres, so that a variability in strength is expected.

This situation can be treated on a statistical basis. The approach, pioneered by Weibull (1951), involves conceptually dividing a length L of fibre into a number of incremental lengths, ΔL_1, ΔL_2, etc. When a stress σ is applied, the parameter n_σ defines the number of flaws per unit length sufficient to cause failure under this stress. The fibre fractures when it has at least one incremental element with such a flaw and for this reason the analysis is often known as a *Weakest Link Theory* (WLT). The probability of any given element failing depends on n_σ and on the length of the element. For the first element,

$$P_{f1} = n_\sigma \Delta L_1 \qquad (2.5)$$

The probability, P_s, of the entire fibre surviving under this stress is the product of the probabilities of survival of each of the N individual elements which make up the fibre. So,

$$P_s = (1 - P_{f1})(1 - P_{f2})\ldots(1 - P_{fN}) \qquad (2.6)$$

Since the length of the elements can be taken as vanishingly small, the corresponding P_f values must be small. Using the approximation $(1 - x) \approx \exp(-x)$, applicable when $x \ll 1$, leads to

$$P_s = \exp[-(P_{f1} + P_{f2} \ldots + P_{fN})] \qquad (2.7)$$

Substituting from Eqn (2.5), and the corresponding equations for the other elements, gives

$$P_s = \exp[-Ln_\sigma] \qquad (2.8)$$

An expression for n_σ is required if this treatment is to be of any use. Weibull proposed that most experimental data for failure of brittle materials conforms to an equation of the form

$$n_\sigma L_0 = \left(\frac{\sigma}{\sigma_0}\right)^m \qquad (2.9)$$

in which m is usually termed the **Weibull modulus** and σ_0 is a normalising strength, which may for our purposes be taken as the most probable strength expected from a fibre of length L_0. Making this assumption, the probability of failure of a fibre of length L, for an applied stress σ, is

$$P_f = 1 - \exp\left[-\left(\frac{L}{L_0}\right)\left(\frac{\sigma}{\sigma_0}\right)^m\right] \qquad (2.10)$$

The Weibull modulus is an important parameter for characterising the strength distribution exhibited by the fibre (or any other brittle material). If the value of m is large (say > 20), then it can be seen from Eqn (2.9) that stresses even slightly below the normalising value σ_0 would lead to a low probability of failure, while if they were slightly above then a high probability would be predicted. Conversely, a low Weibull modulus (say < 10) would introduce much more uncertainty about the strength of a fibre. In practice, many ceramic materials exhibit Weibull moduli in the range 2–15, representing considerable uncertainty about the stress level at which any given specimen is likely to fail.

To check whether a set of strength values conforms to Eqn (2.10), it is convenient to rearrange the equation into a form where a linear relationship is predicted. This is usually done by taking the logarithm of the probability of survival ($P_s = 1 - P_f$)

$$\ln(P_s) = -\left(\frac{L}{L_0}\right)\left(\frac{\sigma}{\sigma_0}\right)^m \qquad (2.11)$$

so that

$$\ln\left(\frac{1}{P_s}\right) = \left(\frac{L}{L_0}\right)\left(\frac{\sigma}{\sigma_0}\right)^m \tag{2.12}$$

Taking logarithms again then gives

$$\ln\left[\ln\left(\frac{1}{P_s}\right)\right] = \ln(L) - \ln(L_0) + m\ln(\sigma) - m\ln(\sigma_0) \tag{2.13}$$

A plot, in this form, of data for P_s as a function of σ, should give a straight line with a gradient of m. An example is shown in Fig. 2.8, which gives data (Martineau *et al.* 1984) for the strength distributions of three types of SiC monofilament. It can be seen that these data do conform approximately to Eqn (2.13), in that the plots are more or less linear. The two carbon-cored fibres have about the same average strength, but rather different variabilities (m values of 2 and 4). The tungsten-cored fibre, on

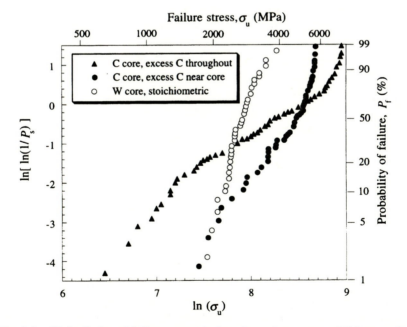

Fig. 2.8   Weibull plot of failure strength data from three types of SiC monofilament, each having been manufactured under different conditions. These data were obtained by testing a fairly large number of individual fibres of each type. The gradients (Weibull moduli m) of the three plots are about 2, 4 and 8. (From Martineau *et al.* 1984).

the other hand, has a lower average strength, but much lower variability ($m = 8$). These differences can be attributed to the nature and distribution of the flaws which are present.

The variability of strength exhibited by most ceramic fibres has important consequences for the mechanical behaviour of composite materials. It means, for example, that points of fibre fracture are often fairly randomly distributed and do not necessarily become concentrated in a single crack plane which propagates through the material. This leads to wide distributions of damage and promotes fibre pull-out (see §9.2.4), enhancing the toughness.

2.3 Matrices

The properties exhibited by various types of matrix are presented in Table 2.5. Information of this type, when considered together with data for reinforcements, immediately allows potential systems to be appraised. For example, glass is evidently of no use for reinforcement of metals if enhancement of stiffness is a primary aim. Slightly more subtle points, such as whether fibre and matrix have widely differing thermal expansion coefficients (and would hence be prone to differential thermal contraction stresses), may also be explored. In practice, however, ease of manufacture (see Chapter 11) often assumes considerable importance. In the next three sections some points are made concerning the factors which affect the choice of matrix.

2.3.1 Polymer matrices

(i) *Thermosetting resins.* The most commonly used resins are epoxy, unsaturated polyester and vinyl ester. These cover a very broad class of chemicals and a wide range of physical and mechanical properties can be obtained. In thermosetting polymers, the liquid resin is converted into a hard rigid solid by chemical cross-linking, which leads to the formation of a tightly bound three-dimensional network. This is usually done while the composite is being formed. The mechanical properties depend on the molecular units making up the network and on the length and density of the cross-links. The former is determined by the initial chemicals used and the latter by control of the cross-linking processes in the cure. Curing can be achieved at room temperature, but it is usual to use a cure schedule which involves heating at one or more temperatures for predetermined times to achieve optimum cross-linking and hence optimum

Table 2.5 Selected properties for different types of matrix

Matrix	Density ρ (Mg m^{-3})	Young's modulus E (GPa)	Poisson's ratio ν	Tensile strength σ_* (GPa)	Failure strain ϵ_* (%)	Thermal expansivity α (10^{-6} K^{-1})	Thermal conductivity K (W m^{-1} K^{-1})
Thermosets							
epoxy resins	1.1–1.4	3–6	0.38–0.40	0.035–0.1	1–6	60	0.1
polyesters	1.2–1.5	2.0–4.5	0.37–0.39	0.04–0.09	2	100–200	0.2
Thermoplastics							
Nylon 6.6	1.14	1.4–2.8	0.3	0.06–0.07	40–80	90	0.2
polypropylene	0.90	1.0–1.4	0.3	0.02–0.04	300	110	0.2
PEEK	1.26–1.32	3.6	0.3	0.17	50	47	0.2
Metals							
Al	2.70	70	0.33	0.2–0.6	6–20	24	130–230
Mg	1.80	45	0.35	0.1–0.3	3–10	27	100
Ti	4.5	110	0.36	0.3–1.0	4–12	9	6–22
Ceramics							
borosilicate glass	2.3	64	0.21	0.10	0.2	3	12
SiC	3.4	400	0.20	0.4	0.1	4	50
Al$_2$O$_3$	3.8	380	0.25	0.5	0.1	8	30

properties. A relatively high-temperature final post-cure treatment is often given to minimise any further cure and change in properties during service. Shrinkage during cure and thermal contraction on cooling after cure can lead to residual stresses in the composite.

It can be seen from the data in Table 2.5 that thermosets have slightly different properties from thermoplastics. Notable among these are much lower strains to failure. Thermosets are essentially brittle materials, while thermoplastics can undergo appreciable plastic deformation. However, there are also significant differences between different types of thermoset. For example, epoxies are in general tougher than unsaturated polyesters or vinyl esters, particularly in the cases of recent advanced epoxy formulations (Monteiro 1986). Further differences between different matrices are highlighted by the data in Table 2.6. This shows, for example, that epoxies can have good resistance to heat distortion and also that they shrink less during curing than polyesters. This is a significant advantage, as is the fact that they can be partially cured, so that pre-pregs can be supplied (see Chapter 11). In fact, epoxies are superior in most respects to the alternative thermosetting systems, which are sometimes preferred simply on grounds of lower cost.

(ii) *Thermoplastics.* Unlike thermosetting resins, thermoplastics are not cross-linked. They derive their strength and stiffness from the inherent properties of the monomer units and the very high molecular weight. This ensures that in amorphous thermoplastics there is a high concentration of molecular entanglements, which act like cross-links, and that in semicrystalline materials there is a high degree of molecular order and alignment. Heating of amorphous materials leads to disentanglement and a change from a rigid solid to a viscous liquid. In crystalline materials heating results in melting of the crystalline phase to give an amorphous viscous liquid. Both amorphous and semi-crystalline polymers may have anisotropic properties, depending on the conditions during solidification. In amorphous polymers this is due to molecular alignment which occurs during melt flow in moulding or subsequently during plastic deformation. Similarly, in crystalline polymers, the crystalline lamellar units can develop a preferred orientation due, for example, to non-uniform nucleation at the surfaces of fibres, or in the flowing melt, and preferential growth in some directions because of temperature gradients in the melt.

The properties of some thermoplastics have been compared with those of thermosets in Tables 2.5 and 2.6. In addition to their high failure

Table 2.6 Comparison between thermosets and thermoplastics of properties relating to dimensional and environmental stability

Property	Thermosets		Thermoplastics		
	epoxy resins	polyester resins	Nylon 6.6	polypropylene	PEEK
Melting temperature (°C)	—	—	265	164	334
Distortion temperature (°C)	50–200	50–110	120–150	80–120	150–200
Shrinkage on curing (%)	1–2	4–8	—	—	—
Water absorption (24h @ 20 °C) (%)	0.1–0.4	0.1–0.3	1.3	0.03	0.1
Chemical resistance	Good, attacked by strong acids	Attacked by strong acids and alkalis	Good, attacked by strong acids	Excellent	Excellent

strains, they tend to exhibit good resistance to attack by chemicals and generally good thermal stability. This latter point is particularly true of the various advanced thermoplastics developed fairly recently for use in composites. Polyether ether ketone (PEEK), a semi-crystalline polymer, is a good example. The stiffness and strength of this polymer are very little affected by heating up to 150 °C, a temperature at which most polymers have become substantially degraded. A composite containing 60 vol. % of carbon fibres in a PEEK matrix, designated APC–2 by the manufacturers (ICI), has found extensive use in aerospace applications. Other high-performance thermoplastics include polysulphones, polysulphides, and polyimides (Muzzy and Kays 1984, Cogswell 1992). Most of these are amorphous polymers. Many thermoplastics also show good resistance to absorption of water, although this is not true of the nylons (see Table 2.6), which usually have a high degree of crystallinity. All thermoplastics yield and undergo large deformations before final fracture and their mechanical properties are strongly dependent on the temperature and applied strain rate. Another important feature of all thermoplastics is that under constant load conditions the strain tends to increase with time, i.e. creep occurs (see §10.2). This means that there may be a redistribution of the load between matrix and fibres during deformation and under in-service loading conditions.

One of the most significant features of thermoplastic composites is that processing tends to be more difficult than with thermosets. This is essentially because they are already polymeric, and hence highly viscous even when liquid, before the composite is fabricated. Although their glass transition and melting temperatures, T_g and T_m, are in many cases quite low, the melts they produce have high viscosities and cannot easily be impregnated into fine arrays of fibres. Usually it is necessary to ensure that flow distances are short, for example by interleaving thin polymer sheets with fibre preforms, and to apply substantial pressures for appreciable times (see §11.1.5). Once fibre and matrix have been brought together in some way, then various shaping operations, such as injection moulding (§11.1.4) can be carried out.

2.3.2 Metal matrices

The development of metal matrix composites has been concentrated on three metals, *aluminium*, *magnesium* and *titanium*. Metals are normally alloyed with other elements to improve their physical and mechanical properties and a wide range of alloy compositions is available. Final properties are strongly influenced by thermal and mechanical treatments which determine the microstructure. Some typical properties of common metal matrices are given in Table 2.5. The metals used for composites are usually ductile and essentially isotropic. Unlike polymers, the increases in stiffness achieved by incorporation of the reinforcement are often relatively small. However, important improvements are often achieved in properties such as wear characteristics, creep performance and resistance to thermal distortion (Clyne and Withers 1993). All three metals are very reactive, with a strong affinity for oxygen. This has implications for the production of composites, particularly in regard to chemical reactions at the interface between the matrix and the reinforcement, which has proved especially troublesome for titanium.

2.3.3 Ceramic matrices

Four main classes of ceramic have been used for ceramic matrix composites. *Glass ceramics* are complex glass-forming oxides, such as borosilicates and aluminosilicates which have been heat-treated so that a crystalline phase precipitates to form a fine dispersion in the glassy phase (Phillips 1983). Glass ceramics have lower softening temperatures than crystalline ceramics and are easier to fabricate; this is an especially important consideration for ceramic composites. *Conventional ceramics*, such as SiC, Si_3N_4, Al_2O_3 and ZrO_2 are fully crystalline and have the normal structure of crystalline grains randomly oriented relative to each other. There is interest in layered ceramic structures (Clegg *et al.* 1990), in which sheets produced by a powder route are separated by thin, weak interlayers, such as graphite. No fibres are involved in this case and classification as a composite material is rather marginal. Again, one of the main attractions lies in relative ease of processing. There has been considerable interest in the reinforcement of *cement* and *concrete*, usually adding short fibres in such a way that moulding capabilities are not severely impaired (Hannant 1983). Finally, *carbon/carbon* composites, produced by vapour infiltration of an array of carbon fibres, form a specialised but commercially important sub-class of composite material (Fitzer 1987).

The main advantages of adding ceramic fibres to ceramics (or generating interfaces in some other way) is to improve their toughness. Ceramics are very brittle and even a small increase in toughness may be advantageous. Usually the toughness increase comes from repeated crack deflection at interfaces and the nature of the interface assumes overriding importance.

References and further reading

Carbon fibres
Bennett, S. C. and Johnson, D. J. (1978) Structural heterogeneity in carbon
    fibers, *Proc. 5th London Carbon and Graphite Conf., vol. 1*, Soc. for Chem.
    Ind.: London, pp. 377–86
DelMonte, J. (1981) *Technology of Carbon and Graphite Fiber Composites*. Van
    Nostrand Reinhold: New York
Donnet, J-P. and Bansal, R. C. (1984) *Carbon Fibers*. Dekker: New York
Guigon, M., Oberlin, A. and Desarmot, G. (1984) Microtexture and structure
    of some high-modulus, PAN-based carbon fibers, *Fiber Sci. Technol.*, **20**
    177–98
Hamada, T., Nishida, T., Sajiki, Y., Matsumoto, M. and Endo, M. (1987)
    Structures and physical properties of carbon fibers from coal tar
    mesophase pitch, *J. Mater. Res.*, **2** 850–7
Kowalski, I. M. (1987) New high performance domestically produced carbon
    fibers, *SAMPE J.*, **32** 953–63
Matsumoto, T. (1985) Mesophase pitch and its carbon fibres, *Pure & Appl.
    Chem.*, **57** 1537–41
Moreton, R., Watt, W., and Johnson, W. (1967) Carbon fibres of high strength
    and high breaking strain, *Nature*, **213** 690–1
Oberlin, A., Endo, M. and Koyama, T. (1976) Filamentous growth of carbon
    through benzene decomposition, *J. Cryst. Growth*, **32** 335–49
Otani, S. (1965) On the carbon fiber from the molten pyrolysis product,
    Carbon, **3** 31–8
Tibbetts, G. G., Devour, M. G. and Rodda, E. G. (1987) An adsorption–
    diffusion isotherm and its application to the growth of carbon filaments
    on iron catalyst particles, *Carbon*, **25** 367–75
Watt, W., Phillips, L. N. and Johnson, W. (1966) High strength, high modulus
    carbon fibres, *The Engineer*, **221** 815–16
Weeton, J. W., Peters, D. M. and Thomas, K. L. (1987) *Engineer's Guide to
    Composite Materials*. ASM: Metals Park, Ohio, pp. 5–9
Weibull, W. (1951) A statistical distribution function of wide applicability, *J.
    Appl. Mech.*, **18** 293–305

Glass fibres
Gagin, L. V. (1980) The development of fiberglass – a history of compositions
    and materials, *Can. Clay Ceram.*, **53** 10–14
Mohr, J. G. and Rowe, W. P. (1978) *Fiber Glass*. Van Nostrand Reinhold:
    New York.
Proctor, B. A. (1980) Glass fibres for cement reinforcement, *Phil. Trans. Roy.
    Soc. London*, **294A** 427–36

Organic fibres
Black, W. B. (1980) High modulus/high strength organic fibres, *Ann. Rev.
    Mater. Sci.*, **10** 311–62
Dinwoodie, J. M. (1965) Tensile strength of individual compression wood
    fibres and its influence on properties of paper, *Nature*, **205** 763–4
Dinwoodie, J. M. (1981) *Timber. Its Nature and Behaviour*. Van Nostrand
    Reinhold: New York

Dobb, M. G., Johnson, D. J. and Saville, B. P. (1980) Structural aspects of high modulus aromatic polyamide fibres, *Phil. Trans. Roy. Soc. London*, **294A** 483–5

Jones, R. S. and Jaffe, M. (1985) High performance aramid fibres, in *High Technology Fibers*. M. Lewin and J. Preston (eds.), Dekker: New York

Marsh, P. (1990) Breaking the mould, *New Scientist*, June 9, 58–60

Other reinforcements

Abis, S. (1989) Characteristics of an aluminium alloy/alumina metal matrix composite, *Comp. Sci. Tech*, **35** 1–11

Birchall, J. D. (1983) The preparation and properties of polycrystalline aluminium oxide fibres, *J. Brit. Ceram. Soc.*, **83** 143–5

Birchall, J. D., Stanley, D. R., Mockford, M. J., Pigott, G. H. and Pinto, P. J. (1988) The toxicity of silicon carbide whiskers, *J. Mater. Sci. Letts.*, **7** 350–2

Clyne, T. W., Bader, M. G., Cappleman, G. R. and Hubert, P. A. (1985) The use of a δ-alumina fibre for metal matrix composites, *J. Mater. Sci.*, **20** 85–96

Dhingra, A. K. (1980a) Alumina fiber FP, *Phil. Trans. Roy. Soc. London*, **294A** 411–17

Dhingra, A. K. (1980b) Metal matrix composites reinforced with FP fiber, *Phil. Trans. Roy. Soc. London*, **294A** 559–64

Le Petitcorps, Y., Lahaye, M., Pailler, R. and Naslain, R. (1988) Modern boron and SiC CVD filaments; a comparative study, *Comp. Sci. & Tech.*, **32** 31–55

Levitt, A. P. (ed.) (1970) *Whisker Technology*. Wiley: New York

Martineau, P., Lahaye, M., Pailler, R., Naslain, R., Couzi, M. and Cruege, F. (1984) SiC filament/titanium matrix composites regarded as model composites, *J. Mater. Sci.*, **19** 2731–70

McDanels, D. L. (1985) Analysis of stress–strain, fracture and ductility of aluminum matrix composites containing discontinuous silicon carbide reinforcement, *Metall. Trans.*, **16A** 1105–15

Okamura, K., Sato, M. and Hasegawa, Y. (1987) Silicon nitride fibre and silicon oxynitride fibre obtained by the nitridation of polycarbosilane, *Ceram. Int.*, **13** 55–61

Simon, G. and Bunsell, A. R. (1984) Mechanical and structural characterisation of the Nicalon[TM] SiC fibre, *J. Mater. Sci.*, **19**, 3649–57

Yajima, S., Hasegawa, Y., Hayashi, J. and Limura, M. (1978) Synthesis of continuous SiC fibres with high tensile strength and high Young's modulus, *J. Mater. Sci.*, **13** 2569–76

Yajima, S., Okamura, K., Hayashi, J. and Omori, M. (1976) Synthesis of continuous SiC fibres with high tensile strength, *J. Am. Ceram. Soc.*, **59** 324–6

Composite matrices

Bruins, P. F. (ed.) (1976) *Unsaturated Polyester Technology*. Gordon & Breach: New York

Clegg, W. J., Kendall, K., Alford, N. M., Birchall, D. and Button, T. W. (1990) A simple way to make tough ceramics, *Nature*, **347** 455–7

Clyne, T. W. and Withers, P. J. (1993) *An Introduction to Metal Matrix Composites*. Cambridge University Press: Cambridge

Cogswell, F. N. (1992) *Thermoplastic Aromatic Polymer Composites.*
  Butterworth–Heinemann: Oxford
Fitzer, E. (1987) The future of carbon–carbon composites, *Carbon,* **25**, 163–90
Gillham, J. K. (1987) Formation and properties of network polymeric
  materials, *Polym. Eng. Sci.,* **19** 676–82
Hannant, D. J. (1983) Fibre reinforced cements, in *Handbook of Composites,
  vol. 4 – Fabrication of Composites.* A. Kelly and S. T. Mileiko (eds.)
  Elsevier: New York, pp. 429–500
Monteiro, H. A. (1986) Matrix systems for advanced composites, *Pop. Plast.,*
  31 20–7
Muzzy, J. D. and Kays, A. O. (1984) Thermoplastic versus thermosetting
  structural composites, *Polym. Comp.,* **5** 169–72
Phillips, D. C. (1983) Fibre reinforced ceramics, in *Handbook of Composites,
  vol. 4 – Fabrication of Composites.* A. Kelly and S. T. Mileiko (eds.),
  Elsevier: New York, pp. 373–429
Woodhams, R. T. (1985) History and development of engineering resins,
  Polym. Eng. Sci., **25**, 446–52

3
Fibre architecture

Many composite properties are strongly dependent on the arrangement and distribution of fibres: the fibre architecture. This expression encompasses intrinsic features of the fibres, such as their diameter and length, as well as the volume fraction of fibres and their alignment and packing arrangement. In this chapter, geometrical aspects of fibre architecture are described for both continuous- and short-fibre materials. The starting point is a description of fibre arrangements in laminae (sheets containing aligned long fibres) and the laminates that are built up from these. Other forms of continuous fibre systems, such as woven, knitted and braided materials, are then briefly considered. Arrangements in short-fibre systems are more complex and methods of characterising them are described. Finally, reference is made to a topic closely related to fibre architecture, namely the alignment of short fibres in flowing viscous media. This is relevant to many of the processing techniques described in Chapter 11.

3.1 General considerations

3.1.1 Volume fraction and weight fraction

Although most calculations on composite materials are based on the volume fractions of the constituents, it is sometimes important, particularly when calculating the density of the composite, to use weight fractions. The appropriate conversion equations are:

$$f = \frac{w\rho_f}{w\rho_f + w_m\rho_m} \tag{3.1}$$

and

$$w = \frac{f \rho_f}{f \rho_f + f_m \rho_m} \tag{3.2}$$

where f, f_m are the volume fractions, w, w_m are the weight fractions and ρ_f, ρ_m are the densities of the fibre and matrix respectively.

3.1.2 Fibre packing arrangements

In a unidirectional lamina, Fig. 3.1(a), all the fibres are aligned parallel to each other. Ideally, the fibres can be considered to be arranged on a hexagonal or square lattice as shown in Fig. 3.1(b), with each fibre having a circular cross-section and the same diameter. In fact, glass, aramid and many ceramic fibres do have circular cross-sections with a smooth surface finish, but many carbon fibres, although roughly circular, may have irregular surfaces. There is often a considerable variation in fibre diameter for all types of fibre.

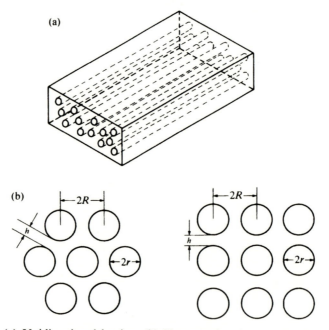

Fig. 3.1   (a) Unidirectional lamina. (b) Hexagonal and square packing of unidirectional fibres.

For the ideal arrangements shown in Fig. 3.1(b), the volume fraction of fibres, f, is related to the fibre radius r by:

$$f = \frac{\pi}{2\sqrt{3}} \left(\frac{r}{R}\right)^2 \qquad \text{(hexagonal)} \qquad (3.3)$$

$$f = \frac{\pi}{4} \left(\frac{r}{R}\right)^2 \qquad \text{(square)} \qquad (3.4)$$

where $2R$ is the closest centre-to-centre spacing of the fibres. The maximum value of f will occur when the fibres are touching, i.e. $r = R$. For a hexagonal array $f_{max} = 0.907$ and for a square array $f_{max} = 0.785$. The separation of the fibres, h, varies with f as:

$$h = 2\left[\left(\frac{\pi}{2f\sqrt{3}}\right)^{1/2} - 1\right]r \qquad \text{(hexagonal)} \qquad (3.5)$$

$$h = 2\left[\left(\frac{\pi}{4f}\right)^{1/2} - 1\right]r \qquad \text{(square)} \qquad (3.6)$$

These equations are represented graphically in Fig. 3.2. Even at low $f(\sim 0.3)$, the closest distance between the fibres is less than one fibre diameter and at higher values of $f(\sim 0.7)$, the spacing becomes very small. This is important when the presence of the fibre modifies the surrounding matrix, since it means that a relatively large proportion of

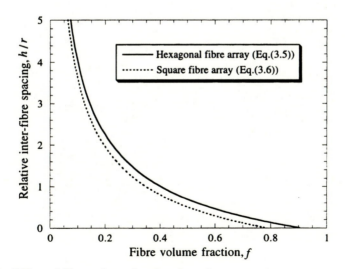

Fig. 3.2   Effect of fibre volume fraction f on the spacing between fibres.

the matrix may be modified, so that it has different properties from the bulk matrix. It is also worth noting that the matrix is often highly constrained between closely spaced fibres and that a highly uniform distribution of fibres is required if physical contact between them is to be avoided (see below).

3.1.3 Clustering of fibres and particles

Experimental studies of the distribution of fibres in unidirectional laminae show that these ideal distributions do not occur in practice, except in small localised regions. An example is given in Fig. 3.3 of a section cut normal to the fibre direction in a lamina with a high fibre content. In some regions the packing closely approximates to an hexagonal array, but there are also matrix-rich regions with irregular packing. Some points of fibre contact are apparent. In laminae with lower fibre contents the packing is often very irregular, with fibre bunching and large matrix-rich regions. Misalignment of the fibres is also much more likely.

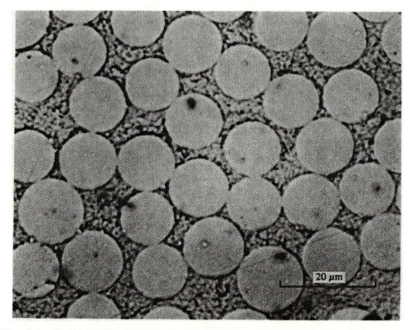

Fig. 3.3 Optical micrograph of a section cut at right angles to fibres in unidirectional laminae of glass fibre/polyester resin.

One of the main consequences of non-regular packing is the difficulty of achieving volume fractions greater than 0.7 and this value must be regarded as the practical limit for commercial materials. It follows that laminae cannot be regarded as being homogeneous from a microstructural point of view, although for the prediction of laminate properties it is assumed that each lamina has a set of characteristic properties.

Another point to note about fibre spacings concerns ease of composite manufacture. Most composite manufacturing operations involve penetration of liquid matrix into an array of fibres (see Chapter 11). When fibres are very closely spaced, initial penetration of liquid involves the generation of sharp meniscus curvature, which requires the application of a high pressure. Furthermore, subsequent flow through the fibre array at an acceptable rate needs a high pressure gradient, particularly if the melt has a high viscosity (as with thermoplastic polymers). The pressures required during processing can therefore become prohibitively high when fibres are closely spaced, particularly if the fibres are fine.

3.2 Long fibres

3.2.1 Laminates

High-performance polymer components usually consist of layers or *laminae* stacked in a pre-determined arrangement. For the prediction of elastic properties of the component as a whole, each lamina may be regarded as homogeneous in the sense that the fibre arrangement and volume fraction are uniform throughout. The fibres in the laminae may be continuous or in short lengths and can be aligned in one or more directions or randomly distributed in two or three dimensions. Two simple arrangements of laminae are illustrated in Fig. 3.4. A unidirectional lamina is often called a **ply** and a stack of laminae is called a **laminate**. The flat laminate in Fig. 3.4(a) consists of identical unidirectional laminae or plies stacked with adjacent plies at 90° to each other. This construction is typical, though considerably simplified, of the material used for high-stiffness panels in aircraft. The curved laminate in Fig. 3.4(b) is part of the wall of a cylindrical vessel. This laminate configuration is commonly found in applications such as pressure pipes and torsion tubes. In this example, the inner lamina is a layer of chopped-strand mat and the outer unidirectional laminae are arranged with the fibres oriented at ± 55° to the axis of the cylinder. Some of the factors affecting the choice of stacking sequence are described in Chapter 5.

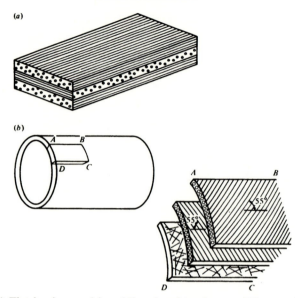

Fig. 3.4   (a) Flat laminate with unidirectional laminae at 90° to each other.
(b) Cylindrical laminate with one layer of chopped-strand mat and two unidirectional laminae.

A simple convention is often used when describing stacking sequences.
This is illustrated by the simple cross-ply laminate shown in Fig. 3.5(a).
The stacking sequence, in relation to the *x*-direction, is written as
$0°/90°/0°/0°/90°/0°$, which can be simplified to $[0/90/0_2/90/0]$, where
the subscript 2 indicates that there are two plies in the 0° orientation.
Since, in this case, the stacking is symmetrical about the mid-plane, the
notation is further simplified to $[0/90/0]_s$ the subscript s denoting that the
stacking sequence is repeated symmetrically. Similarly, the angle-ply
laminate shown in Fig. 3.5(b) is denoted by $[0/+60/-60_2/+60/0]$, which
is abbreviated to $[0/+60/-60]_s$ or $[0/\pm60]_s$.

When the plies do not have the same thickness, or are made of different
materials, it is necessary to specify both the material and the thickness of
each layer, as well as the orientation of the fibres. Thus, the notation $[90/$
$0/2R_c)/(2R_c/0/90)/R_{c0.75}]$ refers to a laminate with fibres in the 0° and 90°
directions, combined with layers of random mat reinforcement (R_c)
which have two different thicknesses, R_c and $0.75 R_c$. Similarly, the
notation Kevlar^{TM} 49/T300 carbon/Kevlar^{TM} 49, $[0_3/\pm45/90]_s$ refers
to a symmetrical laminate with three plies of Kevlar^{TM} 49 fibres, one
ply of $+45°$ and one ply of $-45°$ T300 carbon fibres and one ply of
90° Kevlar^{TM} 49 fibres. In addition to specifying the ply orientation in

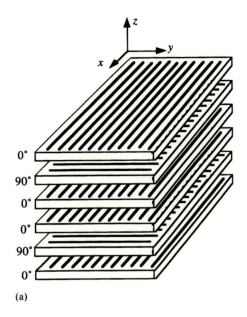

(a)

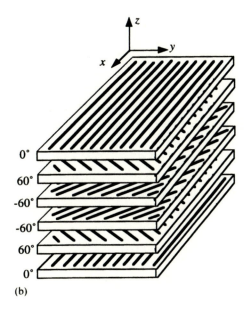

(b)

Fig. 3.5 Arrangement of plies in (a) a crossply laminate and (b) an angle-ply laminate sandwiched between 0° plies.

relation to a reference direction (*x*-direction in Fig. 3.5), it is also necessary, for non-symmetrical laminates, to relate the stacking sequence to
the form of the component. Thus, for the simple example in Fig. 3.4, the
random fibres are on the inside of the pipe, so that the stacking sequence
is: $_{outside}[+55/-55/R_c]_{inside}$.

3.2.2 Woven, braided and knitted fibre arrays

Continuous fibres can be produced in a variety of geometrical forms, in
addition to stacks of unidirectional plies, using technology originally
developed for textile processes: *weaving, braiding* and *knitting*. The
arrangement of fibres in a woven cloth is illustrated in Fig. 3.6. In woven
cloth, the angle between the warp and weft directions is 90°. The flexibility
of cloth allows draping and shaping to occur, facilitating use in non-planar
structures. The angle between the warp and weft directions will depend on
the extent of drape. A complete characterisation of woven roving composites requires details of weave spacing, number of fibres in each roving,
angle between warp and weft directions and the ratio of the number of
fibres in these directions. Woven structures lead to pockets of matrix at the
cross-over points (see Fig. 3.6(b)) and the maximum fibre content for
woven roving composites is less than for fully aligned materials.

The fibre arrangements produced by two-dimensional braiding are
similar to woven fabrics. Braiding is commonly used for flexible tubes,
with the fibre tows interlacing orthogonally. Stretching such a tube by
increasing the length or the diameter results in rotation of the fibre tows.
More complex shapes can be generated, with the fibre tows meeting at
different angles. Braiding is now used to produce three-dimensional fibre
architectures. As with woven fabric reinforcement, matrix-rich regions
are unavoidable, so the maximum fibre content is less than in laminates
made by stacking unidirectional plies.

Knitting is also used to produce fabric preforms. The fibres are usually
in the form of a staple yarn to facilitate knitting. Many knitting configurations are possible. The yarn is arranged in a repeating series of inter-
meshed loops, so that the orientation of the fibres is changing
continuously in three dimensions. The volume fraction of fibres is relatively low and large matrix pockets cannot be avoided. Novel methods of
laying additional straight fibres in knitted structures, to increase the
properties in specific directions, are being developed.

A commonly used form of fibre distribution, particularly for low-cost
applications, is ***chopped strand mat***. Bundles of relatively long fibres are

Fig. 3.6 (a) SEM micrograph of a woven roving before infiltration with resin.
(b) Photomicrograph of a polished section through a woven roving laminate
parallel to one set of fibres.

assembled together with random in-plane orientations, as shown in Fig. 3.7. The material is easy to handle as a preform and the resultant composite material has isotropic in-plane properties. However, the fibre volume fraction is limited to relatively low values.

3.2.3 *Characterisation of fibre orientations in a plane*

The orientation distributions of the fibres in assemblies such as those in Figs. 3.6 and 3.7 are simple to describe. However, there may be cases where the distribution is more complex. These can be represented using normalised histograms of the type shown in Fig. 3.8. The plots are obtained by recording the orientation of individual fibres, with respect to some reference direction. This can be done automatically by image analysis of a photograph, such as a simple optical or electron micrograph, or an X-ray radiograph. The directions are then divided into a convenient number of 'bins'. For the plots shown, there are 18 of these, at intervals of 10°. The span needed is only 180° since no distinction is made between the two directions along the axis of a fibre. The radius of each bin in the plot is proportional to the fraction of fibres with orientations in

Fig. 3.7 SEM micrograph of chopped-strand mat before infiltration with resin.
(From Darlington *et al.* 1976).

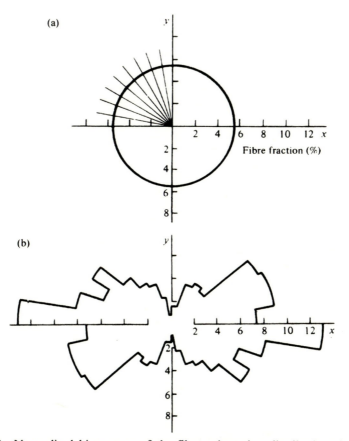

Fig. 3.8 Normalised histograms of the fibre orientation distributions in two-dimensional arrays of fibres. (a) Completely random (isotropic); and (b) an experimental distribution (Darlington *et al.* 1976) from an injection moulded glass fibre/polypropylene composite.

the range concerned. The radii sum to 100%. For the simple isotropic distribution shown in Fig. 3.8(a), the radius of the circle is $100/18 \sim 5.5\%$. For the distribution shown in Fig. 3.8(b), the fibres are aligned preferentially near an axis in the plane.

3.3 Short fibres

3.3.1 Fibre orientation distributions in three dimensions

Orientation distributions are more difficult to measure and characterise when the fibres do not lie parallel to a single plane, which is frequently the case with relatively short fibres. The simplest method (Vincent and Agassant 1986) is based on the assumption that the fibres are straight cylinders of circular section. A planar, undistorted section of the composite is examined in which each fibre is clearly visible as an ellipse, as illustrated in Fig. 3.9. The block *ABCDEFGH* represents a thin parallel-sided slice of material, which has been cut at a pre-determined position and angle with respect to the reference axes of the component. The orientation of the fibre is defined by the two angles α and β. The angle α can be identified on the section (e.g. *ABCD*) as that between the reference direction (marked as y in this case) and the direction of the major axis of the ellipse – see Fig. 3.9(b). The angle β can be measured in one of two ways. The simplest method involves measuring the aspect ratio (major to minor axis) of the section, from which β is readily obtained

$$\beta = \sin^{-1}\left(\frac{b}{a}\right) \tag{3.7}$$

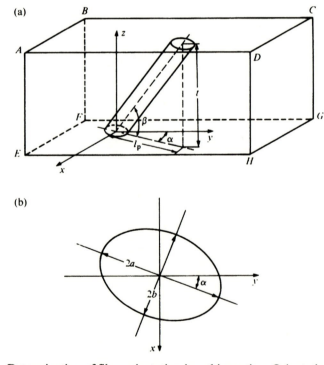

Fig. 3.9   Determination of fibre orientation in a thin section. Orientation defined (a) by angles α and β; and (b) by shape and orientation of fibre cross-section.

However, this ratio may be difficult to measure accurately, particularly if the fibre diameter is small. If the section is transparent (optically or by X-ray radiography), it is also possible to measure the projected length, L_p, of the fibre. The angle β is then given by

$$\beta = \tan^{-1}\left(\frac{t}{L_p}\right) \qquad (3.8)$$

where t is the thickness of the section.

Neither of these methods fully characterises the orientation of the fibre, because there are two possible positions for a fibre having angles α and β. The same aspect ratio and projected length would be obtained from a fibre lying at an angle $(\pi - \beta)$. It may be possible when examining transparent sections to establish which end of the fibre emerges at the top surface, perhaps by comparing optical and X-ray photographs, but this inhibits the automatic acquisition of data by image analysis.

There are other methods of characterising fibre orientation distributions, although their usefulness is rather limited. For example, if an image in which the fibres are opaque and the matrix transparent is derived from a conventional micrograph, then a Fourier transform can be obtained by the diffraction of light passing through this image (Ovland and Kristiansen 1988). If this is repeated for a series of sections, then a full characterisation of the distribution can be obtained. Another possible technique (Juul Jensen *et al.* 1988), applicable only to single-crystal fibres with a known crystallographic direction along the fibre axis (such as SiC whiskers), involves study of diffraction patterns from the component. In many cases, X-rays are not suitable in view of their limited penetration depth (particularly in MMCs) and neutron diffraction is preferable.

Orientation distributions in three dimensions are commonly represented on a *stereographic projection* (stereogram). Thus, *texture* information for polycrystals is presented as *pole figures*, which depict the relative frequencies of the orientation of specified crystallographic directions, relative to the external frame of reference (Bunge 1982, Bunge 1989). Representation of fibre distributions is simpler than for the texture of polycrystals, since only the orientation of the fibre axis is required.

The orientation of each fibre axis is represented as a point on the stereogram. On a stereographic projection, a random (isotropic) three-dimensional distribution of orientations does not plot as a uniform density of points; the points are clustered near the centre and are sparse towards the edges. This is illustrated in Fig. 3.10, which shows (a) how

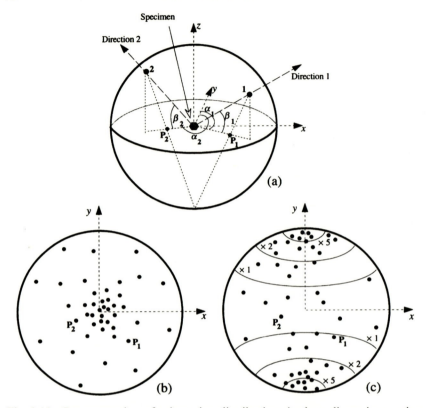

Fig. 3.10   Representation of orientation distributions in three dimensions, using the stereographic projection. (a) Construction of the stereogram, showing how two directions, 1 and 2, are plotted as points P_1 and P_2 where the lines from 1 and 2 to the 'south pole' intersect the 'equatorial plane'. (b) Stereogram of a set of randomly oriented directions. (c) Stereogram of a set of directions, with a systematic bias towards the reference direction y, with superimposed contours separating regions in which the population densities are different multiples of the random case.

two directions, 1 and 2, plot as P_1 and P_2 on the projection and (b) how an isotropic distribution of directions plots as a non-uniform density of points. The most effective way to present fibre distribution information is in the form of a series of contours, representing the ratio of the local density of points to that expected for an isotropic distribution. Such contours are most easily constructed by means of a computer program. An example is shown in Fig. 3.10(c). This also allows the strength of any preferred orientation to be characterised by a single figure, since the value of the highest contour present can readily be established.

3.3.2 Fibre length distributions

Fibre length distribution is important in short-fibre composites. Processing operations such as extrusion, commonly used for thermoplastic and metal matrix composites, can cause extensive fibre fracture, with strong effects on mechanical properties. The techniques used to determine the fibre length distribution can be classified broadly into indirect and direct methods. The indirect methods involve the measurement of some physical property of the composite, such as strength or modulus, which depends on the fibre length. This is an imprecise approach, although it may have some value in quality control. In direct methods, the fibres are separated from the matrix, since it is virtually impossible to make any useful measurements *in situ*. This is usually done by dissolving the matrix, so as to form a suspension of fibres, which are then deposited onto a suitable substrate for examination with the optical or scanning electron microscope (Arsenault 1984). An alternative approach involves filtering the fibres through a series of sieves to separate the different length fractions, but this is subject to errors by long fibres slipping through at steep angles to the sieve plane, or short ones becoming trapped by the presence of other fibres.

Direct measurement is straightforward, but can be time-consuming. An optical micrograph, such as that of Fig. 3.11(a), is examined, either manually or with an image analysis system, to give a series of fibre lengths. The data are plotted as a histogram, with each measured length allocated to a size bin, as shown in Fig. 3.11(b). In this example, there is a pronounced skew to the distribution, with a tail at the long-fibre end. The fibre length before processing was 6 mm and the histogram shows that none of the fibres survived unbroken.

The definition of a meaningful average fibre length is difficult but two simple averages are commonly used. The **number average** fibre length is defined as:

$$L_N = \frac{\sum N_i L_i}{\sum N_i} \tag{3.9}$$

where N_i is the number of fibres of length L_i (i.e. within some specified range near L_i). The **weight (or volume) average** fibre length is defined as:

$$L_W = \frac{\sum W_i L_i}{\sum W_i} \tag{3.10}$$

where W_i is the weight of fibres of length L_i. For fibres of constant diameter, this can be expressed as:

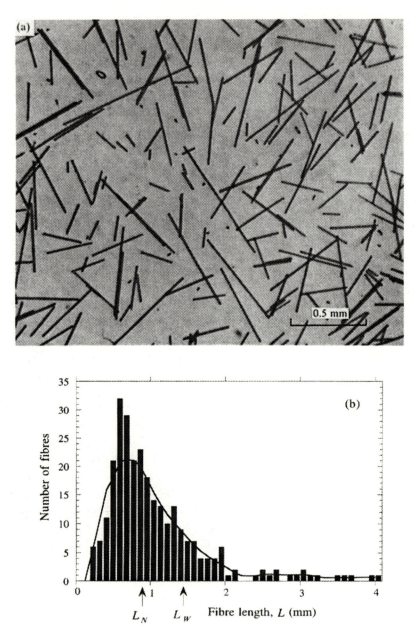

Fig. 3.11   (a) Optical micrograph of short fibres separated from a thermoplastic matrix after an injection process. (b) Length histogram for fibres extracted from a thermoset injection moulding. (Pennington 1979).

$$L_W = \frac{\sum \alpha N_i L_i^2}{\sum \alpha N_i L_i} = \frac{\sum N_i L_i^2}{\sum N_i L_i} \qquad (3.11)$$

where $\alpha = \pi r^2 \rho$ ($2r$ = diameter of fibres, ρ = density). The difference between these two averages is shown in Fig. 3.11(b). The number average L_N is lower than the weight average L_W. The length distribution based on weight is in many ways more meaningful, since it reflects the proportion of the total fibre content with any given length.

3.4 Voids

Several types of voids may be present in composite materials. These can occur in all forms of composite, although there are variations in their incidence depending on fabrication route and matrix type. Large cavities can form during the manufacture of the component as a result of gross defects. Small voids often form adjacent to the fibres, either because of incomplete infiltration during processing or cavitation during deformation. Voids also form in matrix-rich pockets or fibre-free regions between laminae.

There are two main methods of evaluating the void content of materials (including composites). The first is to examine a polished section, identify the voids, either manually or automatically, and determine the area fraction, which is equal to the volume fraction in the absence of any sectioning bias. The sectioning method has the advantage of allowing the locations and shapes of the voids to be established. Examples of several types of voids are shown in Fig. 3.12. The method is often inaccurate, since small voids are difficult to detect and even large ones are easily distorted by flow of matrix or loss of reinforcement fragments during polishing. It is also difficult to establish average void contents in a specimen without examining large numbers of sections.

The second technique, which is free from most of these problems, involves accurate measurement of the density of the sample. Some of the practical points involved in such measurements are discussed by Pratten (1981). The density is determined by weighing the sample in air and then in a liquid of known density. Application of Archimedes' principle leads to the following expression for the density (ρ) of the sample in terms of measured weights (W)

$$\rho = \left(\frac{W_a \rho_L - W_L \rho_a}{W_a - W_L} \right) \qquad (3.12)$$

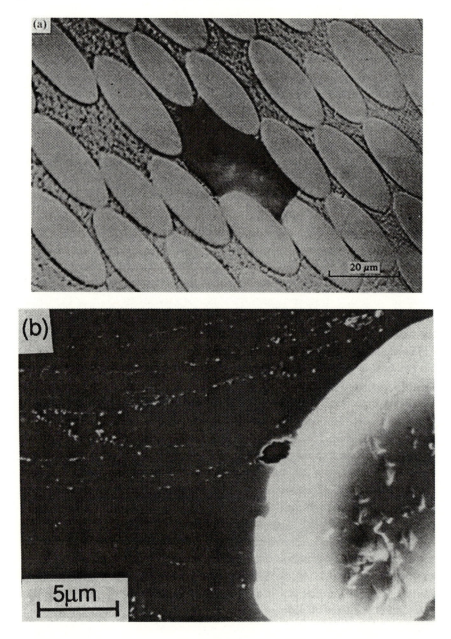

Fig. 3.12 (a) Optical micrograph showing a large void between fibres in a glass fibre/polyester resin lamina. (b) SEM micrograph (Shahani 1991) showing a small cavity adjacent to an alumina particle in an aluminium-based composite after cold drawing.

where the subscripts a and L refer to air and liquid. The liquid should have a high density and chemical stability and a low vapour pressure and surface tension. The most popular liquid currently in use is perfluoro-1-methyl decalin.

A problem arises if any of the pores are surface-connected, since the liquid may then progressively enter the porosity network during weighing, giving variable readings. One solution is to coat the sample with a thin layer of lacquer which is impervious to the liquid. A weighing system with two scale pans in series is used, one above the liquid and one immersed in it. This assembly is weighed (a) with no sample, (b) with the uncoated sample on the upper scale pan, (c) with the coated sample on the upper scale pan and (d) with the coated sample immersed in the liquid on the lower scale pan. The density ρ of the specimen is given from the recorded sample weights by

$$\rho = W_{ua}\left[\left(\frac{W_{ca} - W_{cL}}{\rho_L - \rho_a}\right) - \left(\frac{W_{ca} - W_{ua}}{\rho_c - \rho_a}\right)\right]^{-1} + \rho_a \qquad (3.13)$$

where the subscripts a, L, u and c refer to air, liquid, uncoated and coated, respectively. The density of the coating, ρ_c, must be known, but a very high precision is not essential for this, provided the coating is relatively thin.

This weighing procedure only allows the void content to be found if the density of the fully sound composite is known. This requires that the densities of the constituents and the volume fraction of reinforcement are known accurately. If fibre and matrix have very different densities, then an error in the fibre content significantly affects the calculated void content. In practice, however, the fibre volume fraction can be estimated fairly accurately by metallographic methods. An alternative approach is to compare the measured densities before and after a consolidation operation, such as hot isostatic pressing (HIP).

Finally, ultrasonic scanning (C-scan) can be used for the non-destructive qualitative examination of void distribution and delamination faults in composite materials. The test sample is scanned by an ultrasonic pulse and the attenuation in the material is measured. The information is processed to produced a two-dimensional image of the sample. The method is particularly useful in quality control of sheet materials. An example from a laminated carbon fibre/epoxy resin composite is shown in Fig. 3.13.

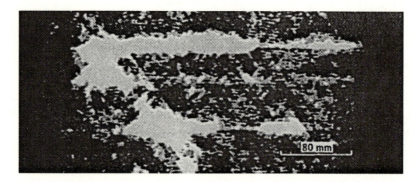

Fig. 3.13 C-scan photograph of a multi-ply carbon fibre epoxy resin laminate. Light regions are due to voids and delaminations in the material.

(a)

(b)

(c)

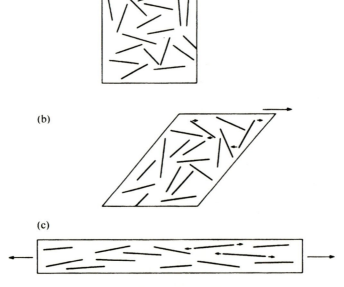

Fig. 3.14 Schematic representation of the changes in fibre orientation occurring during flow. (a) Initial random distribution, (b) rotation during shear flow and (c) alignment during elongational flow.

3.5 Fibre orientation during processing

Changes in fibre orientation often occur during the processing of short-fibre composite materials. In these operations the melt undergoes both elongational or extensional flow and shear flow. An indication of the effect of these flow processes on the fibre orientation is illustrated in Fig. 3.14 for simple two-dimensional deformation. During extensional flow, the fibres rotate towards the direction of the extension. With large extensions, a high degree of alignment can be produced. In shear flow, some of the fibres rotate towards the direction of shear and others rotate in the opposite direction, so that there is no net change in orientation. Thus, the degree of preferred fibre orientation after processing is dependent on the flow field. The viscosity of the matrix affects the final orientation distribution mainly through its effect on the way in which the mould fills. This, in turn, determines the distribution of elongational and shear fields. An example of the effects of flow fields on fibre orientation is given in §11.1.4, which covers the injection moulding process.

References and further reading

Arsenault, R. J. (1984) The strengthening of Al alloy 6061 by fiber and platelet SiC, *Mat. Sci. & Eng.* **64** 171–81

Bunge, H. J. (1982) *Texture Analysis in Materials Science: Mathematical Methods.* Butterworths: New York

Bunge, H. J. (1989) Advantages of neutron diffraction in texture analysis, *Textures and Microstructures*, **10** 265–307

Darlington, M. W., McGinley, P. L. and Smith, G. R. (1976) Structure and anisotropy of stiffness in glass fibre reinforced thermoplastics, *J. Mater. Sci.*, **11** 877–86

Folkes, M. J. and Russell, D. A. M. (1980) Orientation effects during the flow of short fibre reinforced thermoplastics, *Polymer.*, **21** 1252–8

Juul Jensen, D., Lilholt, H. and Withers, P. J. (1988) Determination of fibre orientations in composites with short fibres, in *Mechanical and Physical Behaviour of Metallic and Ceramic Matrix Composites.* S. I. Andersen, H. Lilholt and O. B. Pedersen (eds.) Risø Nat. Lab.: Roskilde, Denmark: pp. 413–20

Ovland, S. and Kristiansen, K. (1988) Characterisation of homogeneity and isotropy of composites by optical diffraction and imaging, in *Mechanical and Physical Behaviour of Metallic and Ceramic Matrix Composites.* S. I. Andersen, H. Lilholt and O. B. Pedersen (eds.) Risø Nat. Lab.: Roskilde, Denmark: pp. 527–32

Pennington, D. (1979) PhD thesis, University of Liverpool

Pratten, N. A. (1981) Review; The precise measurement of the density of small samples, *J. Mater. Sci.*, **16**, 1737–47

Shahani, R. A. (1991) PhD thesis, University of Cambridge

4

Elastic deformation of long-fibre composites

In the previous two chapters, some background was given about the various types of reinforcement and the ways in which they may be distributed within different matrices. In this chapter, attention is turned to the problem of predicting the behaviour of the resulting composites. Prime concern is with the mechanical properties. The incorporation of the reinforcement is usually aimed at enhancing the stiffness and strength of the matrix. The details of this enhancement can be rather complex and difficult to describe with complete rigour. The simplest starting point is to consider the elastic behaviour of a composite with continuous fibres, all aligned in the same direction. Aligned composites are normally used to exploit the stiffness (and strength) parallel to the fibres. However, it is also important to understand the way they behave when loaded in other directions. The treatment therefore includes the behaviour under transverse loading. In this chapter and in the following one, the assumption is made that there is perfect bonding between fibre and matrix across the interface between them. The detailed nature of the interfacial region, and the consequences of imperfect bonding, are considered in Chapter 7.

4.1 Axial stiffness

The simplest treatment of the elastic behaviour of aligned long-fibre composites is based on the premise that the material can be treated as if it were composed of parallel slabs of the two constituents bonded together, with relative thicknesses in proportion to the volume fractions of matrix and fibre. This is illustrated in Fig. 4.1. The two slabs are constrained to have the same lengths parallel to the bonded interface. Thus if a stress is applied in direction of fibre alignment (the 1-direction),

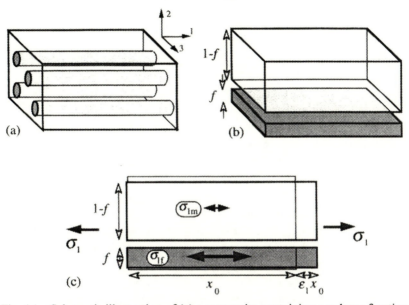

Fig. 4.1 Schematic illustration of (a) a composite containing a volume fraction f of aligned, continuous fibres, and (b) a representation of this as bonded slabs of matrix and fibre material. (c) On applying a stress σ_1 parallel to the fibre axis, the two slabs experience the same axial strain ϵ_1.

both constituents exhibit the same strain in this direction, ϵ_1. This 'equal strain' condition is valid for loading along the fibre axis, provided that there is no interfacial sliding.

It is now a simple matter to derive the Young's modulus of the composite, E_1. The axial strain in the fibre and the matrix must correspond to the ratio between the stress and the Young's modulus for each of the two components, so that

$$\epsilon_1 = \epsilon_{1f} = \frac{\sigma_{1f}}{E_f} = \epsilon_{1m} = \frac{\sigma_{1m}}{E_m} \tag{4.1}$$

Hence, for a composite in which the fibres are much stiffer than the matrix ($E_f \gg E_m$), the reinforcement is subject to much higher stresses ($\sigma_{1f} \gg \sigma_{1m}$) than the matrix and there is a redistribution of the load. The overall stress σ_1 can be expressed in terms of the two contributions being made

$$\sigma_1 = (1-f)\sigma_{1m} + f\sigma_{1f} \tag{4.2}$$

The Young's modulus of the composite can now be written

$$E_1 = \frac{\sigma_1}{\epsilon_1} = \frac{[(1-f)\sigma_{1m} + f\sigma_{1f}]}{(\sigma_{1f}/E_f)} = E_f\left[\frac{(1-f)\sigma_{1m}}{\sigma_{1f}} + f\right]$$

Using the ratio between the stresses in the components given by Eqn (4.1), this simplifies to

$$E_1 = (1-f)E_m + fE_f \qquad (4.3)$$

This well-known 'Rule of Mixtures' indicates that the composite stiffness is simply a weighted mean between the moduli of the two components, depending only on the volume fraction of fibres. This equation is expected to be valid to a high degree of precision, providing the fibres are long enough for the equal strain assumption to apply. (The details of this condition are examined in Chapter 6.) Very minor deviations from the equation are expected as a result of stresses which arise when the Poisson's ratios of the two components are not equal. It may be shown theoretically by means of more advanced treatments, for example, the Eshelby model described in Chapter 6, that the predicted discrepancies are extremely small under all circumstances. Accurate experimental validation of the rule of mixtures has been demonstrated for a number of composites with continuous fibres. The equal strain treatment is often described as a 'Voigt model'.

4.2 Transverse stiffness

Prediction of the transverse stiffness of a composite from the elastic properties of the constituents is far more difficult than the axial value. In addition, experimental measurement of transverse stiffness is more prone to error, partly as a result of higher stresses in the matrix – which can, for example, cause polymeric matrices to creep under modest applied loads. The conventional approach is to assume that the system can again be represented by the 'slab model' depicted in Fig. 4.1. In the fibre composite shown in Fig. 4.1(a), both 2- and 3-directions are transverse to the fibres. An obvious problem with the slab model is that the two transverse directions are not identical; direction 3 is equivalent to the axial direction. In reality, the matrix is subjected to an effective stress intermediate between the full applied stress operating on the matrix when it is normal to the plane of the slab interface and the reduced value calculated in §4.1 for a stress axis parallel to this interface. Before considering this any further, the limiting case of the 'equal stress' model (which is commonly used to describe transverse properties) will be examined. The situation is illustrated schematically in Fig. 4.2. When a stress is applied in the 2-direction

$$\sigma_2 = \sigma_{2f} = \epsilon_{2f} E_f = \sigma_{2m} = \epsilon_{2m} E_m \qquad (4.4)$$

so that the component strains can be expressed in terms of the applied stress. The overall net strain can be written as

$$\epsilon_2 = f\epsilon_{2f} + (1 - f)\epsilon_{2m} \qquad (4.5)$$

from which the composite modulus is given by

$$E_2 = \frac{\sigma_2}{\epsilon_2} = \frac{\sigma_{2f}}{[f\epsilon_{2f} + (1 - f)\epsilon_{2m}]}$$

Substituting expressions for ϵ_{2f} and ϵ_{2m} derived from (4.4) gives

$$E_2 = \left[\frac{f}{E_f} + \frac{(1 - f)}{E_m}\right]^{-1} \qquad (4.6)$$

The equal stress treatment is often described as a 'Reuss model'.

Although this treatment is simple and convenient, it gives a poor approximation for E_2. It is instructive to consider the true nature of the stress and strain distributions during this type of loading when the 'slab' of reinforcement is replaced by fibres. In simple terms, regions of the matrix 'in series' with the fibres, close to them and in line along the loading direction, are subjected to a high stress similar to that carried by the reinforcement – as depicted in Fig. 4.2(b). The regions of the matrix 'in parallel' with the fibres, i.e. adjacent laterally, are constrained to have

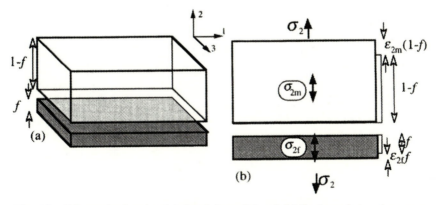

Fig. 4.2   Schematic showing (a) the slab model and (b) the 'equal stress' assumption during transverse stressing.

the same (low) strain as the reinforcement and carry a low stress as illustrated by Fig. 4.1(c).

The overall strain field can be visualised by the operations of removing the fibres, uniformly straining the remaining matrix, re-inserting the fibres (which will be very little deformed by the stress, as they have a high stiffness) and then distorting the matrix so as to re-unite the two components around the interfaces. The result of this operation is shown in Fig. 4.3: the grid lines, which initially form a square mesh to represent unstrained material, become distorted on loading the composite in a way which reveals the distribution of local strain. This strain, and hence the stress, is distributed inhomogeneously within the matrix – in contrast to the uniformity of matrix strain when the loading direction is along the fibres. This inhomogeneity, with sharp concentrations of stress in certain locations, is very significant in terms of the onset of non-elastic behaviour, which arises as a result of interfacial debonding, matrix plastic deformation and microcracking.

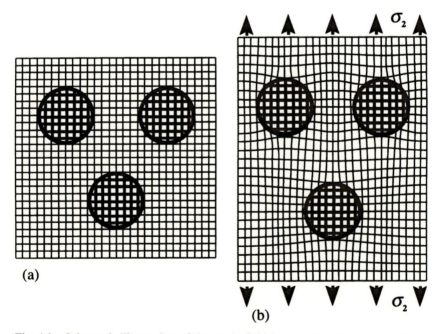

Fig. 4.3 Schematic illustration of the strain field in a composite transverse to an hexagonal array of parallel fibres which are much stiffer than the matrix. (a) Unstrained and (b) on application of a transverse stress in the vertical direction.

The general nature of the transverse strain field can be confirmed by experiment. For example, the technique of photoelasticity provides a convenient method of characterising the elastic strain fields in transparent materials (e.g. see Marloff and Daniel 1969, Withers *et al.* 1991). The photoelastic image shown in Fig. 4.4 is a transverse slice through a macromodel composite material loaded normal to the fibre direction (vertically in the image). The optical birefringence of the material at a particular location is proportional to the difference between the principal strains (and hence, for an elastically isotropic material, to the difference between the principal stresses) at that point. Contours of equal birefringence therefore correspond to regions in which the deviatoric, or shape-changing, component of the strain has the same magnitude. When viewed between crossed polars, dark fringes appear which represent such contours. The higher-order fringes, which in this case are located above and below the fibres, therefore represent regions in which the matrix is highly

Fig. 4.4 A photoelastic image (Puck 1967), showing isochromatic (equal bi-refringence) fringes for a macromodel composite loaded in transverse tension (vertical direction).

distorted. Comparison of Figs. 4.3(b) and 4.4 shows that the expected strain field is broadly consistent with the observed fringe pattern and confirms that the major distortions occur in regions of high matrix tensile stress. The sharp gradients of stress along the loading direction form an important feature: most simple attempts to represent the distribution of matrix stress (see, for example, Chamis 1987 and Spencer 1987) do not take this into account.

The non-uniform distribution of stress and strain during transverse loading means that the simple equal stress model is inadequate. The slab model gives an underestimate of the Young's modulus and can be treated as a lower bound. Various empirical or semi-empirical expressions designed to give more accurate estimates have been proposed. The most successful of these is that due to Halpin and Tsai (1967). This is not based on rigorous elasticity theory, but broadly takes account of enhanced fibre load bearing, relative to the equal stress assumption. Their expression for the transverse stiffness is

$$E_2 = \frac{E_m(1 + \xi \eta f)}{(1 - \eta f)} \qquad (4.7)$$

in which
$$\eta = \frac{\left(\dfrac{E_f}{E_m} - 1\right)}{\left(\dfrac{E_f}{E_m} + \xi\right)}$$

The value of ξ may be taken as an adjustable parameter, but its magnitude is generally of the order of unity. The expression gives the correct values in the limits of $f = 0$ and $f = 1$ and in general gives good agreement with experiment over the complete range of fibre content.

A comparison is presented in Fig. 4.5 between the predictions of Eqns (4.3), (4.6) and (4.7) and experimental data for a glass fibre/polyester system. It is clear that the equal strain treatment (Eqn (4.3)) is in close agreement with data for the axial modulus. For the transverse modulus, the situation is less clear. Firstly, the experimental data show considerable scatter; some of the values actually lie below the equal stress prediction, Eqn (4.6), which should constitute a lower bound. Secondly, many of the values appear to lie closer to the equal stress curve than to the Halpin–Tsai prediction, although this is less obvious for the high fibre contents. This behaviour is almost certainly the result of inelastic deformation of the matrix. These values were obtained by mechanical loading experiments, in which relatively large

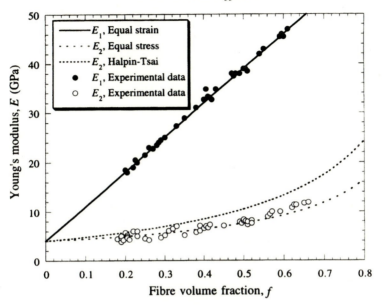

Fig. 4.5 Comparison between experimental data (Brintrup 1975) for the axial and transverse Young's moduli, E_1 and E_2, for polyester/glass fibre composites and corresponding predictions from the equal strain model (Eqn (4.3) for E_1 and the equal stress (Eqn (4.6)) and Halpin–Tsai (Eqn (4.7), with $\xi = 1$) models for E_2. The experimental E_2 values have been affected by inelastic deformation of the matrix.

stresses were present for appreciable times. (This is much less significant during axial loading, since the matrix stresses are so low.) Plastic deformation and creep may occur during transverse testing of this type, particularly with thermoplastic polymers, and this will lead to an underestimate of the true stiffness. In general, tests with stronger and more creep-resistant matrices, or under conditions where all the stresses are kept low and are of short duration (as with dynamic methods of stiffness measurement (Wolfenden and Wolla 1989)), have confirmed that the transverse moduli of long-fibre composites agree quite well with the Halpin–Tsai prediction, Eqn (4.7).

Beyond these simple models for predicting the transverse modulus, there are powerful, but complex, analytical tools such as the Eshelby equivalent homogeneous inclusion approach (see Chapter 6) and numerical techniques such as finite element modelling. The plots shown in Fig. 4.6 give an

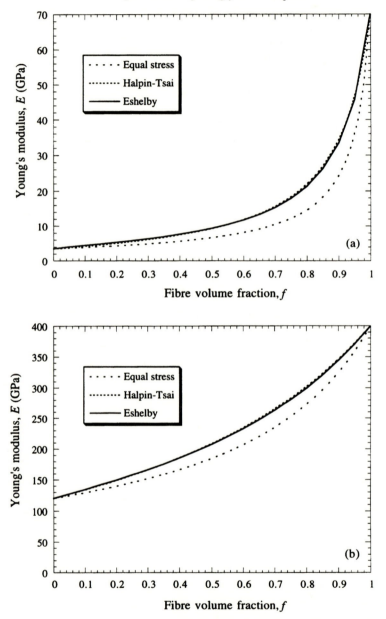

Fig. 4.6 Predicted dependence on fibre volume fraction of the transverse Young's moduli of continuous-fibre composites, according to the equal stress, Eqn (4.6), Halpin–Tsai, Eqn (4.7) and Eshelby models for (a) glass fibres in epoxy and (b) silicon carbide fibres in titanium.

idea of the errors likely to be introduced in real cases by use of simple analytical expressions, as compared with the Eshelby method, which should be more reliable than those from the simpler models. It can be seen that the equal stress assumption gives a significant underestimate for both PMCs and MMCs, which have large and small modulus mismatch respectively. The Halpin–Tsai equation, on the other hand, is quite reliable.

In practice, the behaviour may be influenced by other factors which are difficult to incorporate into simple models. These include the effects of a degree of fibre misalignment, elastic anisotropy of the fibre (or of the matrix – e.g. for a textured polycrystalline metal) or the early onset of a non-elastic response. Nevertheless, it should be noted that, even in the absence of any such complications, use of the equal stress model introduces significant errors: this should be borne in mind, for example, if it is being used in laminate elasticity analysis (see Chapter 5).

4.3 Shear stiffness

The shear moduli of composites can be predicted in a similar way to the axial and transverse stiffnesses, using the slab model. This is done by evaluating the net shear strain induced when a shear stress is applied to the composite, in terms of the individual displacement contributions from the two constituents. It is important to understand the nomenclature convention which is used. A shear stress designated τ_{ij} $(i \neq j)$ refers to a stress acting in the i-direction on the plane with a normal in the j-direction. Similarly, a shear strain γ_{ij} is a rotation towards the i-direction of the j-axis. The shear modulus G_{ij} is the ratio of τ_{ij} to γ_{ij}. As the composite body is not rotating, the condition $\tau_{ij} = \tau_{ji}$ must hold. In addition, $G_{ij} = G_{ji}$ so that $\gamma_{ij} = \gamma_{ji}$. Since the 2- and 3-directions are equivalent in the aligned fibre composite, it follows that there are two shear moduli, because $G_{12} = G_{21} = G_{13} = G_{31} \neq G_{23} = G_{32}$.

There are also two shear moduli for the slab model (Fig. 4.7), but these are unlikely to correspond closely with the values for the fibre composite. The stresses τ_{12} and τ_{21} are assumed to operate equally within both of the constituents. The derivation is similar to the equal stress treatment leading to Eqn (4.6) for transverse stiffness

$$\tau_{12} = \tau_{12f} = \gamma_{12f} G_f = \tau_{12m} = \gamma_{12m} G_m$$

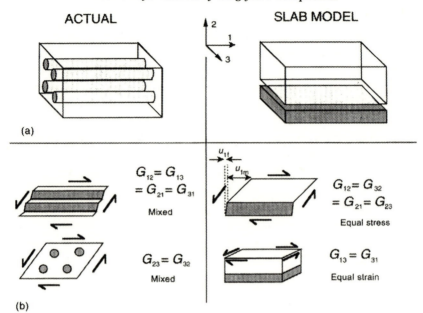

Fig. 4.7 Schematic illustration of how the shear moduli are defined for a real fibre composite and for the slab model representation, indicating how stress and strain partition between the two constituents in each case.

where γ_{12f} and γ_{12m} are the individual shear strains in the two constituents. The total shear strain is found by summing the two contributions to the total shear displacement in the 1-direction

$$\gamma_{12} = \frac{(u_{1f} + u_{1m})}{f + (1 - f)} = f\gamma_{12f} + (1 - f)\gamma_{12m}$$

$$\therefore G_{12} = \frac{\tau_{12}}{\gamma_{12}} = \frac{\tau_{12f}}{f\gamma_{12f} + (1 - f)\gamma_{12m}} = \left[\frac{f}{G_f} + \frac{(1 - f)\gamma_{12m}}{\tau_{12f}}\right]^{-1}$$

i.e. $G_{12} = \left[\dfrac{f}{G_f} + \dfrac{(1 - f)}{G_m}\right]^{-1}$ (4.8)

The other shear modulus shown by the slab model, $G_{13} = G_{31}$ in Fig. 4.7, corresponds to an equal shear strain condition and is analogous to the axial tensile modulus case. It is readily shown that

$$G_{13} = fG_f + (1 - f)G_m \qquad (4.9)$$

which is similar to Eqn (4.3). It may be noted that neither the equal stress condition nor the equal strain condition are close to the situation during shearing of the fibre composite, in which the strain partitions unevenly within the matrix. Therefore neither of the above equations is expected to be very reliable, particularly the equal strain expression.

It is not obvious just how poor the approximation represented by Eqn (4.8) is likely to be, nor even which of the two actual shear moduli it will approach more closely. In fact more rigorous methods predict that the values of G_{12} and G_{23} are rather close to each other, with G_{12} slightly larger in magnitude. Equation (4.8) gives a significant underestimate relative to both of them, while Eqn (4.9) is a gross overestimate. In view of this, the semi-empirical expressions of Halpin and Tsai (1967), mentioned in the last section, are frequently employed. In this case, the appropriate equation is

$$G_{12} = \frac{G_m(1 + \xi\eta f)}{(1 - \eta f)} \qquad (4.10)$$

in which
$$\eta = \frac{\left(\dfrac{G_f}{G_m} - 1\right)}{\left(\dfrac{G_f}{G_m} + \xi\right)}$$

and the parameter ξ is again often taken to have a value of around unity. This has been done for the curves in Fig. 4.8, which shows comparisons between the predictions of Eqn (4.10) and those of the equal stress (Eqn (4.8)) and Eshelby models for both polymer and metal matrix composites. It can be seen that the Halpin–Tsai expression represents a fairly good approximation to the axial shear modulus (G_{12}). A striking feature of both the transverse and the shear moduli for polymer matrix composites (Figs. 4.6(a) and 4.8(a)) is that they are close to the matrix values up to relatively high fibre volume fractions, although in both cases the true modulus is not as low as the prediction of the equal stress model.

4.4 Poisson contraction effects

The Poisson's ratio ν_{ij} describes the contraction in the j-direction on applying a stress in the i-direction and is defined by the equation

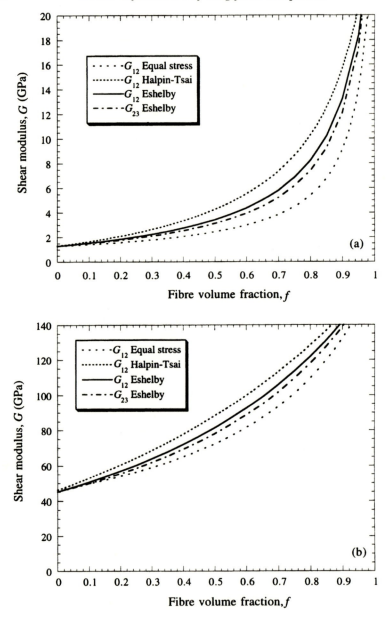

Fig. 4.8   Predicted dependence on fibre volume fraction of the shear moduli of continuous-fibre composites, according to the equal stress equation for G_{12}, Eqn (4.8), the Halpin–Tsai expression for G_{12}, Eqn (4.10) and the Eshelby model (G_{12} and G_{23}). Data are shown for (a) glass fibres in epoxy and (b) silicon carbide fibres in titanium.

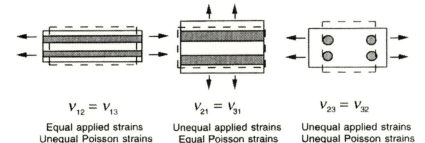

$V_{12} = V_{13}$ $V_{21} = V_{31}$ $V_{23} = V_{32}$

Equal applied strains   Unequal applied strains   Unequal applied strains
Unequal Poisson strains   Equal Poisson strains   Unequal Poisson strains

Fig. 4.9   Schematic illustration of how the three Poisson's ratios are defined for a fibre composite.

$$\nu_{ij} = -\frac{\epsilon_j}{\epsilon_i} \tag{4.11}$$

For an aligned fibre composite, there are three different Poisson's ratios, as illustrated in Fig. 4.9. This brings the total number of elastic constants identified for this material to seven. However, because some of these constants are inter-related, only five independent values are needed to describe the behaviour of such a transversely isotropic material (e.g. see Nye 1985). The following two relationships between identified constants account for this

$$\frac{\nu_{12}}{E_1} = \frac{\nu_{21}}{E_2} \tag{4.12}$$

and

$$G_{23} = \frac{E_2}{2(1 + \nu_{23})} \tag{4.13}$$

Estimation of the ν_{ij} values on the basis of the slab model presents difficulties because of the greater degree to which the contractions of the two constituents must match, when compared with the real composite. The effect of this is that, although three Poisson's ratios can be identified for the slab model, a meaningful calculation can only be done for the equal imposed strain case, giving ν_{12} – see Fig. 4.9. In this case, the Poisson strains for the two constituents can be evaluated independently and summed. Thus

$$\epsilon_{2f} = -\nu_f \epsilon_{1f} = -\nu_f \frac{\sigma_{1f}}{E_f}$$

$$\epsilon_{2m} = -\nu_m \epsilon_{1m} = -\nu_m \frac{\sigma_{1m}}{E_m}$$

so that

$$\epsilon_2 = -\left[\frac{f\nu_f \sigma_{1f}}{E_f} + \frac{(1-f)\nu_m \sigma_{1m}}{E_m}\right]$$

$$= -[f\nu_f \epsilon_1 + (1-f)\nu_m \epsilon_1]$$

and

$$\nu_{12} = -\frac{\epsilon_2}{\epsilon_1} = f\nu_f + (1-f)\nu_m \tag{4.14}$$

A simple rule of mixtures is therefore applicable and, because the equal strain assumption is accurate for axial stressing of the composite, this is expected to be a reliable prediction.

In fact, simple expressions can also be derived to give fairly realistic predictions for the other two ratios. The ratio of the axial contraction to the transverse extension on stressing transversely, ν_{21}, is obtained from the reciprocal relationship given as Eqn (4.12), so that

$$\nu_{21} = [f\nu_f + (1-f)\nu_m]\frac{E_2}{E_1} \tag{4.15}$$

This will be lower than ν_{12} because, on stressing transversely, the fibres will offer strong resistance to axial contraction. This leads to pronounced contraction in the other transverse direction, so that ν_{23} is expected to be high. An expression for ν_{23} may be obtained by considering the overall volume change experienced by the material (Clyne 1990)

$$\Delta = \epsilon_1 + \epsilon_2 + \epsilon_3 = \frac{\sigma_H}{K} \tag{4.16}$$

in which σ_H is the applied hydrostatic stress and K is the bulk modulus of the composite. Only a single stress, σ_2, is being applied here, so that

$$\sigma_H \left(= \frac{\sigma_1 + \sigma_2 + \sigma_3}{3}\right) = \frac{\sigma_2}{3}$$

Thus

$$\epsilon_3 = \frac{\sigma_2}{3K} - \epsilon_1 - \epsilon_2 \tag{4.17}$$

which leads directly to

$$\nu_{23} = -\frac{\epsilon_3}{\epsilon_2} = -\frac{\sigma_2}{3K\epsilon_2} + \frac{\epsilon_1}{\epsilon_2} + 1$$

$$\text{i.e. } \nu_{23} = 1 - \nu_{21} - \frac{E_2}{3K} \tag{4.18}$$

The bulk modulus of the composite can be estimated via an equal stress assumption, which should be quite accurate in this case, so that

$$\sigma_H = \Delta_f K_f = \Delta_m K_m$$

and

$$\Delta = f\Delta_f + (1-f)\Delta_m$$

giving

$$K = \frac{\sigma_H}{\Delta} = \left[\frac{f}{K_f} + \frac{(1-f)}{K_m}\right]^{-1} \tag{4.19}$$

The bulk moduli of the constituents are related to other elastic constants by expressions such as

$$K_f = \frac{E_f}{3(1 - 2\nu_f)}$$

so that it is a simple matter to evaluate ν_{23} from the standard elastic constants of the constituents.

The accuracy of Eqn (4.18) is determined largely by the error in E_2. In the comparisons shown in Fig. 4.10, the Halpin–Tsai values of E_2, predicted by Eqn (4.8), were used in obtaining values for ν_{21} and ν_{23}. It can be seen that agreement with the Eshelby predictions is fairly good. These plots convey an idea of the pronounced tendency under transverse loading for the composite to contract in the other transverse direction in preference to the axial direction. Such effects are of particular significance for the behaviour of laminates (see Chapter 5).

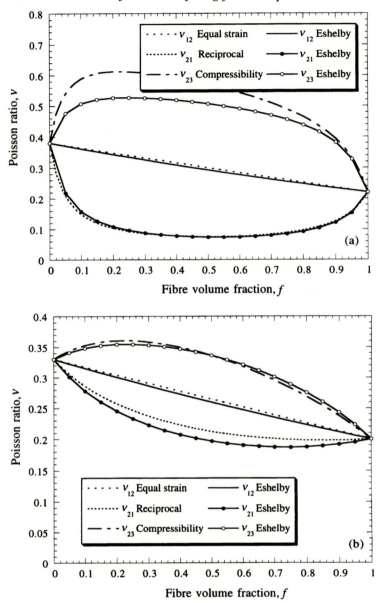

Fig. 4.10 Predicted dependence on fibre volume fraction of the three Poisson's ratios of continuous fibre composites, according to the equal strain condition, Eqn (4.14), for ν_{12}, reciprocal relation, Eqn (4.15), for ν_{21}, compressibility expression, Eqn (4.18), for ν_{23} and the Eshelby model for all three. Predictions are shown for (a) glass fibres in epoxy; and (b) silicon carbide fibres in titanium.

References and further reading

Brintrup, A. (1975) Dr.Ing. thesis, Technische Hochschule: Aachen

Chamis, C. C. (1987) Simplified composite micromechanics for predicting microstresses, *J. Reinf. Plast. & Compos.*, **6** 268–89

Clyne, T. W. (1990) A compressibility-based derivation of simple expressions for the transverse Poisson's ratio and shear modulus of an aligned long fibre composite, *J. Mater. Sci. Letts.*, **9** 336–9

Halpin, J. C. and Tsai, S. W. (1967) Environmental factors in composite design, *Air Force Materials Laboratory Technical Report*, AFML-TR-67-423

Jones, R. M. (1975) *Mechanics of Composite Materials.* McGraw-Hill: New York

Marloff, R. H. and Daniel, I. M. (1969) Three-dimensional photoanalysis of a fibre-reinforced composite model, *Expl Mech.*, **9** 156–62

Nye, J. F. (1985) *Physical Properties of Crystals – Their Representation by Tensors and Matrices.* Clarendon: Oxford

Puck, A. (1967) Zur Beanspruchen und Verformung von GFK-Mehrschichtenverbund-Bauelementen, *Kunstoffe*, **57** 965–73

Spencer, A. (1986) The transverse moduli of fibre composite material, *Comp. Sci. & Techn.*, **27** 93–109

Withers, P. J., Chorley, E. M. and Clyne, T. W. (1991) Use of the frozen stress photoelastic method to explore load partitioning in short fibre composites, *Mat. Sci. & Eng.*, **A135**, 173–8

Wolfenden, A. and Wolla, J. M. (1989) Mechanical damping and dynamic modulus measurements in alumina and tungsten fibre reinforced aluminium, *J. Mater. Sci.*, **24** 3205–12

5

Elastic deformation of laminates

In the last chapter, it was shown that an aligned composite is stiff along the fibre axis, but relatively compliant in the transverse directions. Sometimes, this is all that is required. For example, in a slender beam, such as a fishing rod, the loading is often predominantly axial and transverse or shear stiffness are not important. However, there are many applications in which loading is distributed within a plane: these range from panels of various types to cylindrical pressure vessels. Equal stiffness in all directions within a plane can be produced using a planar random assembly of fibres. This is the basis of chopped-strand mat. However, demanding applications require material with higher fibre volume fractions than can readily be achieved in a planar random array. The approach adopted is to stack and bond together a sequence of thin 'plies' or 'laminae', each composed of long fibres aligned in a single direction, into a laminate. It is important to be able to predict how such a construction responds to an applied load. In this chapter, attention is concentrated on the stress distributions which are created and the elastic deformations which result. This involves consideration of how a single lamina will deform on loading at an arbitrary angle to the fibre direction. A brief summary is given first of some matrix algebra used in elasticity theory.

5.1 Elastic deformation of anisotropic materials

5.1.1 Hooke's law

A review of some basic points about stress and strain is appropriate. The reader is referred to standard textbooks such as Dieter (1986) and Nye (1985) for detailed treatments. The state of stress at a point in a body is defined by the nine components of the **stress tensor** σ_{ij}, in which the stress acts in the *i*-direction on the plane with a normal in the *j*-direction. When $i = j$, σ_{ij} is a normal stress and if $i \neq j$, it represents a shear stress, often written as τ_{ij}. Unless the body is continuously rotating, $\tau_{ij} = \tau_{ji}$, so that there are only six independent components. This is also true of the resulting strain, which is represented by the tensor ε_{ij}.

Care must be taken in distinguishing the **strain tensor** ε_{ij} from the **relative displacement tensor** e_{ij}. Application of a shear stress τ_{ij}, together with τ_{ji}, produces angular rotations of the *i*- and *j*-directions in the body by e_{ji} and e_{ij} respectively. These relative displacements represent a combination of distortion (strain) and rigid body rotation. Limiting cases arise when $e_{ij} = e_{ji}$ (no rotation) and $e_{ij} = -e_{ji}$ (no distortion). Note that e_{ij} is positive when it involves rotating the positive *j*-direction towards the positive *i*-axis. Identification of the strain and rigid body rotation components of e_{ij} involves writing it as the sum of symmetric and antisymmetric tensors

$$e_{ij} = \frac{1}{2}(e_{ij} + e_{ji}) + \frac{1}{2}(e_{ij} - e_{ji}) = \varepsilon_{ij} + \omega_{ij} \qquad (5.1)$$

in which ε_{ij} is the strain tensor and ω_{ij} is the **rotation tensor**. Now, for the normal strains $(i = j)$, there is no scope for confusion because $e_{ii} = \varepsilon_{ii}$, which is the conventional engineering strain, written as ε_i when the axes are principal directions. However, for $i \neq j$, the **engineering shear strain** γ_{ij} differs[†] from the **tensorial shear strain** ε_{ij}

$$\gamma_{ij} = 2\varepsilon_{ij} = e_{ij} + e_{ji} \qquad (5.2)$$

Some examples of the relationships between these angles for different types of shear loading are shown in Fig. 5.1. Careful note should be taken of the factor of 2 relating engineering and tensorial shear strains because, while γ is often more convenient in practical use, tensor operations such as rotation of the axes must be carried out using ε (see below).

The relationship between σ_{ij} and ε_{ij} can be expressed as

$$\sigma_{ij} = C_{ijkl}\varepsilon_{kl} \qquad (5.3)$$

where C_{ijkl} is the **stiffness tensor**. This is a fourth-rank tensor with $3^4 (= 81)$ components and Eqn (5.3) represents 9 equations. For each

[†]The root cause of this is that the engineering shear strain is conventionally related, via the shear modulus, to a single shear stress, whereas in practice the observed value of γ arises from the simultaneous operation of a pair of shear stresses.

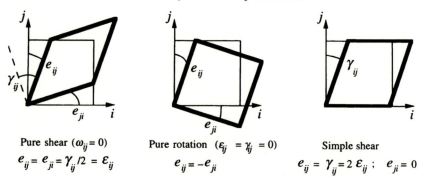

Pure shear $(\omega_{ij} = 0)$ Pure rotation $(\varepsilon_{ij} = \gamma_{ij} = 0)$ Simple shear

$e_{ij} = e_{ji} = \gamma_{ij}/2 = \varepsilon_{ij}$ $e_{ij} = -e_{ji}$ $e_{ij} = \gamma_{ij} = 2\,\varepsilon_{ij}\,;\quad e_{ji} = 0$

Fig. 5.1 Examples (Dieter 1986) of how 2-D displacement components can represent different combinations of shear and rigid body rotation: (a) pure shear, (b) pure rotation and (c) simple shear.

equation (pair of i and j values), the terms which appear are dictated by the ***Einstein summation convention***. This states that when a suffix appears twice in a product then terms are summed with respect to that suffix. The first of these equations would therefore be

$$\sigma_{11} = C_{1111}\varepsilon_{11} + C_{1112}\varepsilon_{12} + C_{1113}\varepsilon_{13}$$
$$+ C_{1121}\varepsilon_{21} + C_{1122}\varepsilon_{22} + C_{1123}\varepsilon_{23}$$
$$+ C_{1131}\varepsilon_{31} + C_{1132}\varepsilon_{32} + C_{1133}\varepsilon_{33}$$

5.1.2 Effects of symmetry

In practice, the equations are often much simpler than these general forms. For example, the symmetry of all the stress and strain tensors when the body is in static equilibrium means that

$$C_{ijkl} = C_{ijlk} = C_{jikl} = C_{jilk} \tag{5.4}$$

which reduces the number of independent constants in C_{ijkl} to 36. In view of this symmetry, a contracted form, termed the ***matrix notation***, can be employed

$$\sigma_p = C_{pq}\varepsilon_q \tag{5.5}$$

in which p and q run from 1 to 6. It can be shown that by converting ij to p and kl to q according to the scheme

tensor notation	11	22	33	23,32	31,13	12,21
matrix notation	1	2	3	4	5	6

Eqn (5.5) may be written out as

$$
\begin{bmatrix} \sigma_1 \\ \sigma_2 \\ \sigma_3 \\ \tau_{23} \\ \tau_{31} \\ \tau_{12} \end{bmatrix} = \begin{bmatrix} C_{11} & C_{12} & C_{13} & C_{14} & C_{15} & C_{16} \\ C_{21} & C_{22} & C_{23} & C_{24} & C_{25} & C_{26} \\ C_{31} & C_{32} & C_{33} & C_{34} & C_{35} & C_{36} \\ C_{41} & C_{42} & C_{43} & C_{44} & C_{45} & C_{46} \\ C_{51} & C_{52} & C_{53} & C_{54} & C_{55} & C_{56} \\ C_{61} & C_{62} & C_{63} & C_{64} & C_{65} & C_{66} \end{bmatrix} \begin{bmatrix} \varepsilon_1 \\ \varepsilon_2 \\ \varepsilon_3 \\ \gamma_{23} \\ \gamma_{31} \\ \gamma_{12} \end{bmatrix}
$$

In practice, it is sometimes more useful to be able to express observed strains in terms of applied stresses, using the **compliance tensor** S_{ijkl}

$$ \varepsilon_{ij} = S_{ijkl}\sigma_{kl} $$

and in this case a similar reduction to matrix notation

$$ \varepsilon_p = S_{pq}\sigma_q \tag{5.6} $$

can be carried out, with additional conversion factors such that

$S_{pq} = S_{ijkl}$ when both p and q are 1, 2 or 3

$S_{pq} = 2S_{ijkl}$ when either p or q is 4, 5 or 6

$S_{pq} = 4S_{ijkl}$ when both p and q are 4, 5 or 6

These factors arise from the relation between engineering and tensorial shear strains. Finally, thermodynamic considerations relating to elastic strain energy lead to

$$ C_{pq} = C_{qp} \tag{5.7} $$
$$ S_{pq} = S_{qp} $$

further reducing the number of independent constants in these matrices from 36 to 21. The reader is referred to Nye (1985) for a clear exposition of this and the conversion from tensor to matrix notation.

The number of independent constants in C_{pq} or S_{pq} is often less than 21, as a result of symmetry exhibited by the material itself. In Fig. 5.2, an indication is given of this number, and inter-relationships between the elements of the S_{pq} or C_{pq} matrices, for several types of material. Note that a single lamina is normally at least **orthotropic**, i.e. having three mutually perpendicular planes of symmetry, and is often **transversely isotropic**. In a material with a high degree of symmetry, the number of independent elastic constants is small. As an example, the version of Eqn (5.6) applicable to an isotropic material is

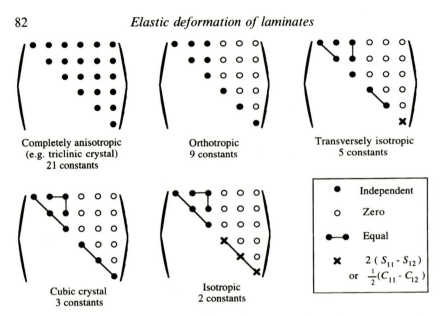

Fig. 5.2   Some examples (Nye 1985) of the form of C_{ij} and S_{ij} matrices for materials exhibiting different types of symmetry. All the matrices are symmetrical about the leading diagonal.

$$\begin{bmatrix} \varepsilon_1 \\ \varepsilon_2 \\ \varepsilon_3 \\ \gamma_{23} \\ \gamma_{31} \\ \gamma_{12} \end{bmatrix} = \begin{bmatrix} S_{11} & S_{12} & S_{12} & 0 & 0 & 0 \\ S_{12} & S_{11} & S_{12} & 0 & 0 & 0 \\ S_{12} & S_{12} & S_{11} & 0 & 0 & 0 \\ 0 & 0 & 0 & 2(S_{11} - S_{12}) & 0 & 0 \\ 0 & 0 & 0 & 0 & 2(S_{11} - S_{12}) & 0 \\ 0 & 0 & 0 & 0 & 0 & 2(S_{11} - S_{12}) \end{bmatrix} \begin{bmatrix} \sigma_1 \\ \sigma_2 \\ \sigma_3 \\ \tau_{23} \\ \tau_{31} \\ \tau_{12} \end{bmatrix} \quad (5.8)$$

so that for an applied stress σ_1

$$\varepsilon_1 = S_{11}\sigma_1$$
$$\varepsilon_2 = \varepsilon_3 = S_{12}\sigma_1$$
$$\gamma_{23} = \gamma_{31} = \gamma_{12} = 0$$

The elements of these matrices can always be expressed in terms of engineering elastic constants and in this case it is evident that

$$S_{11} = \frac{1}{E}$$
$$S_{12} = -\frac{\nu}{E}$$

It can be seen that two elastic constants are sufficient to define fully the behaviour of an isotropic material.

5.2 Off-axis elastic constants of laminae

5.2.1 Calculation procedure

Elementary analyses involve the assumption that each lamina is in a plane stress state, so that $\sigma_3 = \tau_{23} = \tau_{31} = 0$. This is a good approximation for a thin, isolated lamina, but may become somewhat inaccurate for a laminate as a result of through-thickness constraint effects – see §5.4. Eqn (5.6) can now be written in a simplified form, assuming orthotropic symmetry

$$\begin{bmatrix} \varepsilon_1 \\ \varepsilon_2 \\ \gamma_{12} \end{bmatrix} = [S] \begin{bmatrix} \sigma_1 \\ \sigma_2 \\ \tau_{12} \end{bmatrix} = \begin{bmatrix} S_{11} & S_{12} & 0 \\ S_{12} & S_{22} & 0 \\ 0 & 0 & S_{66} \end{bmatrix} \begin{bmatrix} \sigma_1 \\ \sigma_2 \\ \tau_{12} \end{bmatrix} \qquad (5.9)$$

in which, by inspection of the individual equations, it can be seen that

$$S_{11} = \frac{1}{E_1}$$

$$S_{12} = -\frac{\nu_{12}}{E_1} = -\frac{\nu_{21}}{E_2}$$

$$S_{22} = \frac{1}{E_2}$$

$$S_{66} = \frac{1}{G_{12}}$$

Similar consideration of Eqn (5.5) leads to

$$\begin{bmatrix} \sigma_1 \\ \sigma_2 \\ \tau_{12} \end{bmatrix} = \begin{bmatrix} C_{11} & C_{12} & 0 \\ C_{12} & C_{22} & 0 \\ 0 & 0 & C_{66} \end{bmatrix} \begin{bmatrix} \varepsilon_1 \\ \varepsilon_2 \\ \gamma_{12} \end{bmatrix} \qquad (5.10)$$

in which

$$C_{11} = \frac{E_1}{1 - \nu_{12}\nu_{21}}$$

$$C_{12} = \frac{\nu_{12}E_2}{1 - \nu_{12}\nu_{21}} = \frac{\nu_{21}E_1}{1 - \nu_{12}\nu_{21}}$$

$$C_{22} = \frac{E_2}{1 - \nu_{12}\nu_{21}}$$

$$C_{66} = G_{12}$$

Note that the relations expected with transverse isotropy (see Fig. 5.2)

$$S_{66} = 2(S_{11} - S_{12})$$

$$C_{66} = \frac{1}{2}(C_{11} - C_{12})$$

are *not* applicable here, because exclusion of the through-thickness direction eliminates this symmetry from the lamina. As a result, there are *four* independent elastic constants. An immediate point to note about Eqns (5.9) and (5.10) is that there is no **interaction** between normal and shear behaviour; a normal stress gives rise only to normal strains and a shear stress produces only shear strains.

However, this is not the case when the lamina is loaded in some arbi-

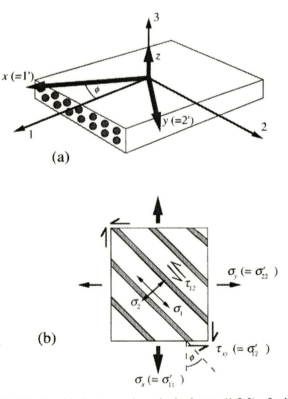

(a)

(b)

Fig. 5.3   (a) Relationship between the principal axes (1,2,3) of a lamina and the coordinate system (x, y, z) for an arbitrary in-plane applied stress. (b) Illustration of how an applied stress system σ'_{ij} (σ_x, σ_y and τ_{xy}) produces stresses in the lamina σ_{ij} (σ_1, σ_2 and τ_{12}).

trary direction within the plane. The situation is illustrated in Fig. 5.3. The first step in determining the strains within the lamina is to establish the induced stresses[†], referred to the fibre axis (σ_1, σ_2 and τ_{12}), in terms of the externally applied stress system (σ_x, σ_y and τ_{xy}). This can be done by means of a geometrical construction involving resolution of forces, but it is simpler to use the equation expressing any second rank tensor with respect to a new coordinate frame

$$\sigma_{ij} = a_{ik}a_{jl}\sigma'_{kl} \tag{5.11}$$

in which a_{ik} is the direction cosine of the (new) i-direction referred to the (old) k-direction. Obviously, the conversion will work in either direction, provided the direction cosines are defined correctly. For example, the normal stress parallel to the fibre direction, σ_{11}, sometimes written as σ_1, can be expressed in terms of the applied stresses σ'_{11} ($=\sigma_x$), σ'_{22} ($=\sigma_y$) and σ'_{12} ($=\tau_{xy}$)

$$\sigma_{11} = a_{11}a_{11}\sigma'_{11} + a_{11}a_{12}\sigma'_{12}$$
$$+ a_{12}a_{11}\sigma'_{21} + a_{12}a_{12}\sigma'_{22}$$

Let ϕ be the angle between the fibre axis (1) and the stress axis (x). Then, referring to Fig. 5.3(a), these direction cosines take the values

$$a_{11} = \cos\phi \qquad\qquad a_{12} = \cos(90 - \phi) = \sin\phi$$
$$a_{21} = \cos(90 + \phi) = -\sin\phi \qquad a_{22} = \cos\phi$$

By carrying out this operation for all three stresses it will be seen that

$$\begin{bmatrix} \sigma_1 \\ \sigma_2 \\ \tau_{12} \end{bmatrix} = [T] \begin{bmatrix} \sigma_x \\ \sigma_y \\ \tau_{xy} \end{bmatrix} \tag{5.12}$$

where

$$[T] = \begin{bmatrix} c^2 & s^2 & 2cs \\ s^2 & c^2 & -2cs \\ -cs & cs & c^2 - s^2 \end{bmatrix} \tag{5.13}$$

in which $c = \cos\phi$ and $s = \sin\phi$. The same matrix can be used to transform tensorial strains, so that

[†]Note that these stresses parallel and normal to the fibre axis will not now, in general, be the principal stresses in the system. This can cause confusion in view of the common convention of denoting the principal stresses by σ_1, σ_2 and σ_3.

$$\begin{bmatrix} \varepsilon_1 \\ \varepsilon_2 \\ \varepsilon_{12} \end{bmatrix} = [T] \begin{bmatrix} \varepsilon_x \\ \varepsilon_y \\ \varepsilon_{xy} \end{bmatrix} \tag{5.14}$$

However, to work in terms of engineering strains, using $\gamma_{xy} = 2\varepsilon_{xy}$, etc., (see §5.1), then $[T]$ must be modified (by halving the elements t_{13} and t_{23} and doubling elements t_{31} and t_{32} of the matrix T) so as to give

$$\begin{bmatrix} \varepsilon_1 \\ \varepsilon_2 \\ \gamma_{12} \end{bmatrix} = [T'] \begin{bmatrix} \varepsilon_x \\ \varepsilon_y \\ \gamma_{xy} \end{bmatrix} \tag{5.15}$$

in which

$$[T'] = \begin{bmatrix} c^2 & s^2 & cs \\ s^2 & c^2 & -cs \\ -2cs & 2cs & c^2 - s^2 \end{bmatrix} \tag{5.16}$$

The procedure is now a straightforward progression from the stress–strain relationship when the lamina is loaded along its principal axes, Eqn (5.9), to a general one involving a **transformed compliance tensor** $\overline{S}$, which will depend on ϕ. The first step is to write the inverse of Eqn (5.15), giving the strains relative to the loading direction, i.e. the information required, in terms of the strains relative to the fibre direction. This involves using the inverse of the matrix $[T']$, written as $[T']^{-1}$

$$\begin{bmatrix} \varepsilon_x \\ \varepsilon_y \\ \gamma_{xy} \end{bmatrix} = [T']^{-1} \begin{bmatrix} \varepsilon_1 \\ \varepsilon_2 \\ \gamma_{12} \end{bmatrix} \tag{5.17}$$

in which

$$[T']^{-1} = \begin{bmatrix} c^2 & s^2 & -cs \\ s^2 & c^2 & cs \\ 2cs & -2cs & c^2 - s^2 \end{bmatrix}$$

Now, the strains relative to the fibre direction can be expressed in terms of the stresses in those directions via the on-axis stress–strain relationship for the lamina, Eqn (5.9), giving

$$\begin{bmatrix} \varepsilon_x \\ \varepsilon_y \\ \gamma_{xy} \end{bmatrix} = [T']^{-1}[S] \begin{bmatrix} \sigma_1 \\ \sigma_2 \\ \tau_{12} \end{bmatrix}$$

Finally, the original transform matrix of Eqn (5.12) can be used to express these stresses in terms of those being externally applied, to give the result

$$\begin{bmatrix} \varepsilon_x \\ \varepsilon_y \\ \gamma_{xy} \end{bmatrix} = [T']^{-1}[S][T] \begin{bmatrix} \sigma_x \\ \sigma_y \\ \tau_{xy} \end{bmatrix} = [\overline{S}] \begin{bmatrix} \sigma_x \\ \sigma_y \\ \tau_{xy} \end{bmatrix} \qquad (5.18)$$

The elements of $[\overline{S}]$ are obtained by **concatenation** (the equivalent of multiplication) of the matrices $[T']^{-1}$, $[S]$ and $[T]$. The following expressions are obtained

$$\begin{aligned}
\overline{S}_{11} &= S_{11}c^4 + S_{22}s^4 + (2S_{12} + S_{66})c^2s^2 \\
\overline{S}_{12} &= S_{12}(c^4 + s^4) + (S_{11} + S_{22} - S_{66})c^2s^2 \\
\overline{S}_{22} &= S_{11}s^4 + S_{22}c^4 + (2S_{12} + S_{66})c^2s^2 \\
\overline{S}_{16} &= (2S_{11} - 2S_{12} - S_{66})c^3s - (2S_{22} - 2S_{12} - S_{66})cs^3 \\
\overline{S}_{26} &= (2S_{11} - 2S_{12} - S_{66})cs^3 - (2S_{22} - 2S_{12} - S_{66})c^3s \\
\overline{S}_{66} &= (4S_{11} + 4S_{22} - 8S_{12} - 2S_{66})c^2s^2 + S_{66}(c^4 + s^4)
\end{aligned} \qquad (5.19)$$

It will be seen that $[\overline{S}] \rightarrow [S]$ as $\phi \rightarrow 0$. The behaviour of the lamina is still fully described by four independent elastic constants, since these six elements can all be expressed in terms of S_{11}, S_{12}, S_{22} and S_{66}. A similar procedure can be used to derive the elements of $[\overline{C}]$, the *transformed stiffness tensor*

$$\begin{aligned}
\overline{C}_{11} &= C_{11}c^4 + C_{22}s^4 + (2C_{12} + 4C_{66})c^2s^2 \\
\overline{C}_{12} &= C_{12}(c^4 + s^4) + (C_{11} + C_{22} - 4C_{66})c^2s^2 \\
\overline{C}_{22} &= C_{11}s^4 + C_{22}c^4 + (2C_{12} + 4C_{66})c^2s^2 \\
\overline{C}_{16} &= (C_{11} - C_{12} - 2C_{66})c^3s - (C_{22} - C_{12} - 2C_{66})cs^3 \\
\overline{C}_{26} &= (C_{11} - C_{12} - 2C_{66})cs^3 - (C_{22} - C_{12} - 2C_{66})c^3s \\
\overline{C}_{66} &= (C_{11} + C_{22} - 2C_{12} - 2C_{66})c^2s^2 + C_{66}(c^4 + s^4)
\end{aligned} \qquad (5.20)$$

5.2.2 Engineering constants

Either of these matrices fully defines the elastic response of the material. However, it is often more convenient to represent these characteristics in terms of the conventional engineering constants. These can be obtained from the stiffness or compliance tensors by inspection of the relationships presented as Eqns (5.9) and (5.10). The relationships are simpler if the compliance tensor is used. Thus

$$E_x = \frac{1}{\overline{S}_{11}} \qquad (5.21)$$

$$E_y = \frac{1}{\overline{S}_{22}} \qquad (5.22)$$

$$G_{xy} = \frac{1}{\overline{S}_{66}} \qquad (5.23)$$

$$\nu_{xy} = -E_x \overline{S}_{12} \qquad (5.24)$$

$$\nu_{yx} = -E_y \overline{S}_{12} \qquad (5.25)$$

These engineering constants can therefore be found once the compliance tensor has been evaluated.

Some examples of the behaviour predicted by Eqns (5.19) for two different composites are illustrated in Figs. 5.4, 5.5 and 5.7. The dependence of the Young's and shear moduli of the lamina on the value of ϕ is shown in Fig. 5.4(a) for a polymer matrix composite, using both the equal stress (Eqns (4.6) and (4.8)) and Halpin–Tsai (Eqns (4.7) and (4.10) with $\xi = 1$) expressions for the transverse Young's modulus ($\phi = 90°$) of the composite. The equal stress assumption introduces quite significant errors over a wide range of loading angle, although the predictions do not differ in qualitative terms. The tensile stiffness (Young's modulus) remains close to the theoretical maximum if the stress axis is within a few degrees of the fibre axis, but if ϕ is more than about 5° then it decreases rapidly. The reduction is less pronounced for the metal matrix composite (Fig. 5.4(b)). These predictions have been confirmed by experiments.

The shear stiffness is less sensitive than the Young's modulus to ϕ, but a pronounced peak is always exhibited at 45°. This efficiency of stiff diagonal (45°) members in resisting shear forces is important in many engineering situations – see, for example, the discussions in Gordon (1978).

As mentioned earlier, an important feature of the off-axis loading of unidirectional laminae is the appearance of non-zero 'interaction' terms ($\overline{S}_{16}$ and $\overline{S}_{26}$), indicating that normal stresses produce shear strains and shear stresses produce normal strains. It is convenient to introduce two other engineering constants to characterise the strength of this interaction effect

$$\eta_{xyx} = E_x \overline{S}_{16} \qquad (5.26)$$

$$\eta_{xyy} = E_y \overline{S}_{26} \qquad (5.27)$$

The parameter η_{xyx} therefore represents the ratio of the shear strain (γ_{xy}), induced by the application of a normal stress (σ_x), to the normal strain (ε_x) induced by the same normal stress. It indicates the nature of the

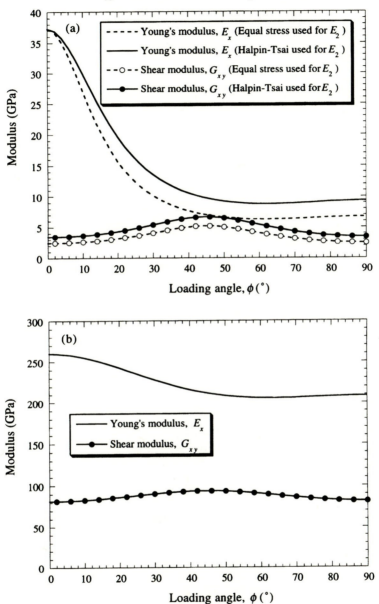

Fig. 5.4 Variation with loading angle ϕ of the Young's modulus E_x and shear modulus G_{xy} for laminae of (a) epoxy/50% glass and (b) titanium/50% SiC monofilament. The transverse Young's moduli were obtained by using the equal stress (Eqn (4.6)) and Halpin–Tsai (Eqn (4.7)) models in (a) and the Halpin–Tsai model only in (b).

Elastic deformation of laminates

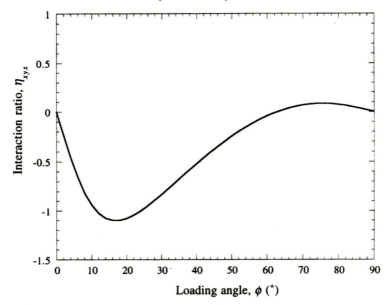

Fig. 5.5 Variation with loading angle ϕ of the interaction ratio η_{xyx} for laminae of epoxy containing 50% of glass fibres, obtained using the Halpin–Tsai expression for E_2.

tensile–shear interaction, although it should be borne in mind that its value is not only proportional to the shear strain observed for a given applied normal stress, but also becomes larger as the tensile stiffness in the loading direction increases. The parameters η_{xyx} and η_{xyy} are often termed *interaction ratios*.

As an example, Fig. 5.5 shows the dependence of η_{xyx} on ϕ. Several features are apparent here. Firstly, substantial changes occur as the loading angle is changed. As expected from simple symmetry arguments, the interaction term is zero at $\phi = 0°$ and $\phi = 90°$, but at intermediate angles the effect can be pronounced. For some values of ϕ, the interaction shear strain is similar in magnitude to the direct normal strain ($\eta_{xyx} \sim -1$). Using the equal stress assumption for E_2 gives a slight overestimate of the interaction term.

These characteristics are of considerable practical significance and can be illustrated using a simple model composite. Fig. 5.6(a) is a photograph of a practical demonstration macromodel. This is made up of four separate laminae, each composed of aligned metal rods in a polyurethane rubber matrix. These metal 'fibres' are oriented so that the stress axis

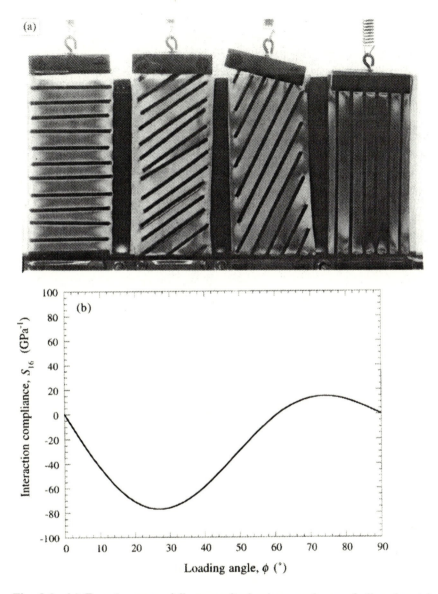

Fig. 5.6 (a) Four 'macromodel' composite laminae made up of aligned metal rods in a polyurethane rubber matrix, each with $f = 0.05$ and the same dimensions when unstressed. All are being subjected to a tensile stress, with the fibre axis making loading angles of $0°$, $30°$, $60°$ and $90°$ to the stress axis. The model is viewed between crossed polars, so that the matrix appears light when it is subject to large shear strains. (b) Predicted variation with loading angle ϕ of the interaction compliance $\bar{S}_{16}$, for the case of the composite material in (a). The Halpin–Tsai expression was used for the transverse modulus E_2.

forms angles of 0°, 30°, 60° and 90° to the fibre direction. There is a strong shear distortion for the 30° lamina, but this is negligible for the others. This is consistent with the plot in Fig. 5.6(b), which shows the variation in the interaction compliance, $\overline{S}_{16}$, with loading angle for the model composite. The observed shear strain is directly proportional to this parameter. It can be seen that the plot is consistent with the behaviour of the model, in that a large shear strain is expected at 30°, but not for the other three orientations. It may also be noted that the degree to which the laminae become extended is broadly consistent with theory; for example, the 60° lamina is predicted to have a slightly lower Young's modulus than the 90° lamina and does indeed appear to become marginally more extended.

The key to a physical interpretation of these effects lies in recognising the role of matrix shear. The matrix in the models in Fig. 5.6(a) are photoelastic and they show that significant matrix shear strains occur in the inclined cases. These shear strains are also responsible for the effects illustrated in Fig. 5.7, which shows that the Poisson's ratio ν_{xy} peaks at an intermediate angle. Large shear strains in the

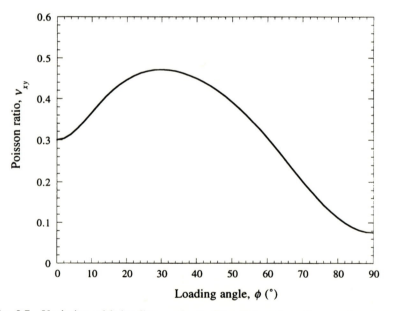

Fig. 5.7   Variation with loading angle ϕ of the Poisson's ratio ν_{xy} of an epoxy/50% glass lamina, obtained using the Halpin–Tsai expression for the transverse modulus E_2.

matrix can induce both shear distortions and large lateral contractions. Effects such as these are important in understanding the behaviour of laminates.

5.3 Elastic deformation of laminates

5.3.1 Loading of a stack of plies

It is evident from Fig. 5.4 that individual laminae containing aligned fibres tend to exhibit highly anisotropic elastic properties. This aniso-tropy can be reduced by stacking a number of laminae (or plies) with different fibre orientations and bonding them together to form a lami-nate. The elastic properties of such a laminate can be predicted from those of the component plies, provided the assembly is taken to be flat and thin, with no through-thickness stresses, and edge effects are neglected. (These are sometimes termed the **Kirchhoff assumptions**.) Detailed analysis of the stress states in laminates can be complex, but the simple treatment outlined below has proved useful for a range of practical situations. Further details are available elsewhere (Jones 1975, Chou and Kelly 1980, Chou 1989).

Calculation of the elastic constants of the laminate follows the scheme illustrated in Fig. 5.8. Note that the loading angle between the stress axis (*x*-direction) and the reference direction for the orientation of the plies (i.e. the $\phi = 0°$ direction) is now expressed as Φ. The fibre direction for the *k*th ply therefore lies at an angle $(\phi - \Phi)$ to the stress axis. The average stress in the *x*-direction can be written both as a function of the stresses in the individual plies and in terms of the overall stiffnesses and strains

$$\sigma_{xg} = \frac{\sum_{k=1}^{n}(\sigma_{xk}t_k)}{\sum_{k=1}^{n}t_k} = \overline{C}_{11g}\varepsilon_{xg} + \overline{C}_{12g}\varepsilon_{yg} + \overline{C}_{16g}\gamma_{xyg} \qquad (5.28)$$

in which t_k is the thickness of the *k*th ply and the subscript g refers to a global value for the whole laminate. As the in-plane strains are con-strained to be the same for all the laminae, the stress in any lamina can be written

$$\sigma_{xk} = \overline{C}_{11k}\varepsilon_{xg} + \overline{C}_{12k}\varepsilon_{yg} + \overline{C}_{16k}\gamma_{xyg}$$

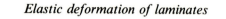

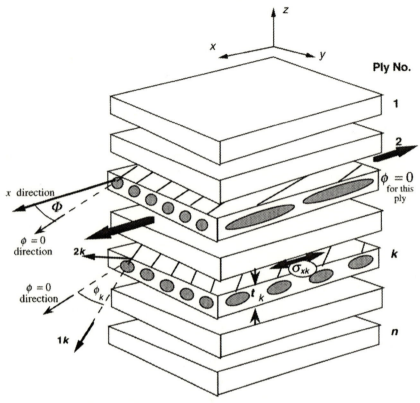

Fig. 5.8 Schematic depicting the loading angle Φ between the x-direction (stress axis) and the reference direction ($\phi = 0°$) for a laminate composed of n plies. Also shown is the angle ϕ_k between the reference direction and the fibre axis of the kth ply.

Substituting this, and equating the coefficients of ε_{xg}

$$\overline{C}_{11g} = \frac{\displaystyle\sum_{k=1}^{n}(\overline{C}_{11k}t_k)}{\displaystyle\sum_{k=1}^{n}t_k} \tag{5.29}$$

Similar expressions apply for the other components. If the corresponding compliances are required, they must be obtained after this operation by application of the inversion relationships

$$S_{11} = \frac{(C_{22}C_{66} - C_{26}^2)}{\Delta}$$

$$S_{22} = \frac{(C_{11}C_{66} - C_{16}^2)}{\Delta}$$

$$S_{12} = \frac{(C_{16}C_{26} - C_{12}C_{66})}{\Delta}$$

$$S_{66} = \frac{(C_{11}C_{22} - C_{12}^2)}{\Delta} \qquad (5.30)$$

$$S_{16} = \frac{(C_{12}C_{26} - C_{22}C_{16})}{\Delta}$$

$$S_{26} = \frac{(C_{12}C_{16} - C_{11}C_{26})}{\Delta}$$

in which

$$\Delta = C_{11}C_{22}C_{66} + 2C_{12}C_{16}C_{26} - C_{22}C_{16}^2 - C_{66}C_{12}^2 - C_{11}C_{26}^2$$

5.3.2 *Predicted behaviour*

Plots of Young's modulus $(E_x = 1/\bar{S}_{11g})$ and Poisson's ratio $(\nu_{xy} = -\bar{S}_{12g}E_x)$ against loading angle for different laminate stacking sequences are shown in Fig. 5.9 (a) and (b), respectively. These plots were obtained by repeated application of Eqn (5.29) for each loading angle value, using the equal stress assumption for transverse modulus. Although this is somewhat inaccurate, its use here will not affect the trends being identified. The 0/90 laminate is less anisotropic than the unidirectional lamina. If further plies are introduced, covering the range of orientations with close, even spacings, then the anisotropy becomes smaller and may disappear, so that the stiffness is the same when loaded in any direction. For the case shown in Fig. 5.9(a), this stiffness is about 16.5 GPa. It may be noted that this limiting uniform stiffness cannot be obtained from a simple expression derived via an integration procedure.

Data such as those in Fig. 5.9 indicate that it is not necessary to space the ply orientations very closely in order to achieve in-plane isotropic elastic properties. On the other hand, 0/90 ('crossply') laminates retain pronounced anisotropy. Other lay-up sequences, in which differences in orientation between successive plies differ in magnitude (e.g. $-\theta/+\theta$ – i.e. 'angle-ply' laminates), are also possible – see Chapter 3. In general, prediction of elastic properties is quite complex for an

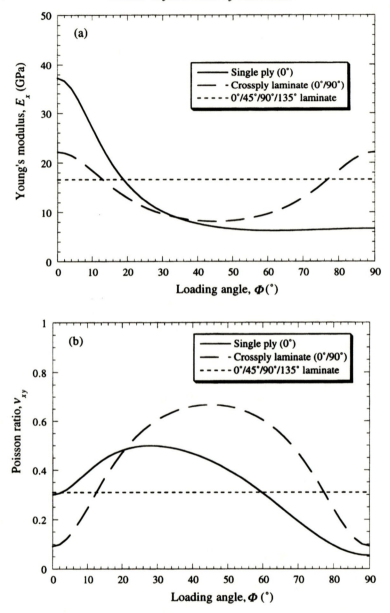

Fig. 5.9 Variation with loading angle Φ of (a) Young's modulus E_x and (b) Poisson's ratio ν_{xy} ($= -S_{12}E_x$) for a single lamina; and for two laminates with different stacking sequences, composed of epoxy/50% glass fibres. (The equal stress model was used for the transverse Young's modulus of a lamina.)

arbitrary stacking sequence, but this operation would normally be carried out by simply inputting the necessary data into a standard computer program. These are now freely available for the main types of personal computer and are in widespread, routine use. It is not necessary to understand the details of the mathematical techniques involved in order to use these programs.

5.4 Stresses and distortions

External loading of a laminate produces complex stress distributions in the individual laminae and between laminae. These may cause shape distortions of the laminate and local microstructural damage and failure. Where possible, it is important to select stacking sequences to minimise undesirable effects such as the distortions which are produced by the tensile–shear interactions described in §5.3.

5.4.1 Balanced laminates

The tensile–shear interaction effects in the laminate as a whole depend on the set of ply orientations. The variation of the interaction ratio, η_{xyx}, with the angle Φ between the stress axis and the reference direction of the laminate ($\phi = 0°$ direction) is shown in Fig. 5.10 for several laminates. As expected, the term becomes small for all Φ as the set of fibre directions become more evenly and more closely spaced, i.e. as the set exhibits greater *rotational symmetry* about the axis normal to the plane of the laminate. In fact, this tensile–shear term (and η_{xyy}) become zero – and the other terms in $\overline{S}_{ijg}$ become constant – for all Φ when there is an N-fold ($N \geq 6$) axis of symmetry, i.e. the fibre directions are evenly spaced at intervals of $360°/N$. (Note that N must be even, as the sense of the fibre direction is immaterial.) The 0/60/120 case shown in Fig. 5.10 corresponds to $N = 6$.

When the tensile–shear interaction terms contributed by the individual laminae all cancel each other out in this way, the laminate is said to be '*balanced*'. Simple crossply and angle-ply laminates are *not* balanced for a general loading angle, although both will be balanced when loaded at $\phi = 0°$ (i.e. parallel to one of the plies for a crossply or equally inclined to the $+\theta$ and $-\theta$ plies for the angle-ply case). If the plies vary in thickness, or in the volume fractions or type of fibres they contain, then even a laminate in which the stacking sequence does exhibit the necessary rotational symmetry is prone to tensile–shear distortions and computation is necessary to determine the lay-up sequence required to construct a

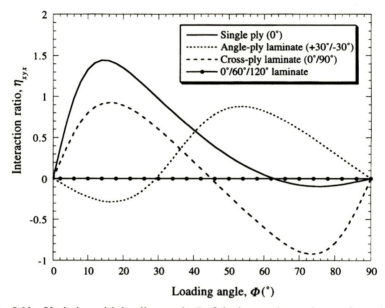

Fig. 5.10   Variation with loading angle Φ of the interaction ratio η_{xyx} for a single lamina and for three laminates with different stacking sequences, composed of epoxy/50% glass fibres. (The equal stress model was used for the transverse Young's modulus of a lamina.)

balanced laminate. In all cases, the stacking order in which the plies are assembled does not enter into these calculations, although it may be important in determining the interlaminar coupling stresses (see below).

5.4.2 Stresses in individual plies of a laminate

The internal stresses may be subdivided into **in-plane** stresses, which can be calculated by the methods outlined below, and **interlaminar** or **through-thickness** stresses, which arise as a result of constraint effects and are more difficult to quantify. The approach to establishing in-plane stresses follows directly from the preceding treatments. Referring to Fig. 5.8, and assuming the stack of laminae to be subject to a stress state σ_x, σ_y and τ_{xy}, the laminate strains are established from the transformed compliance tensor of the laminate.

$$\begin{bmatrix} \varepsilon_x \\ \varepsilon_y \\ \gamma_{xy} \end{bmatrix} = [\overline{S}_{\mathrm{g}}] \begin{bmatrix} \sigma_x \\ \sigma_y \\ \tau_{xy} \end{bmatrix} \qquad (5.31)$$

As these strains are imposed on all the plies, the strains within the kth ply (referred to the fibre direction of the ply, i.e. the $1k$-direction) can be determined from Eqn (5.15), which becomes

$$\begin{bmatrix} \varepsilon_{1k} \\ \varepsilon_{2k} \\ \gamma_{12k} \end{bmatrix} = \left[T' \right]_{\phi = \phi_k} \begin{bmatrix} \varepsilon_x \\ \varepsilon_y \\ \gamma_{xy} \end{bmatrix} \tag{5.32}$$

The stresses in the ply are then obtained from these strains, using the (on-axis) stiffness matrix for the kth ply

$$\begin{bmatrix} \sigma_{1k} \\ \sigma_{2k} \\ \tau_{12k} \end{bmatrix} = \left[C \right]_k \begin{bmatrix} \varepsilon_{1k} \\ \varepsilon_{2k} \\ \gamma_{12k} \end{bmatrix} \tag{[(5.10)]}$$

The stresses in the kth ply are related directly to the applied stresses by

$$\begin{bmatrix} \sigma_{1k} \\ \sigma_{2k} \\ \tau_{12k} \end{bmatrix} = \left[C \right]_k \left[T' \right]_{\phi = \phi_k} \left[\overline{S}_g \right] \begin{bmatrix} \sigma_x \\ \sigma_y \\ \tau_{xy} \end{bmatrix} \tag{5.33}$$

Some results from such a calculation are shown in Fig. 5.11, which gives the variation of σ_{1k}, as a ratio to the applied stress, with the angle Φ

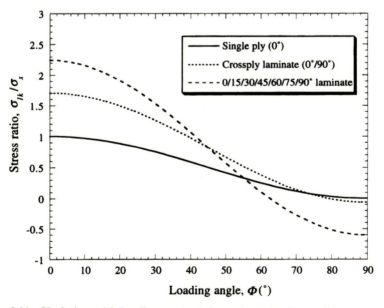

Fig. 5.11 Variation with loading angle Φ for a single lamina and for two laminates of epoxy/50% glass fibres, of the stress σ_{1k} parallel to the fibre axis in a ply initially oriented at 0° to the stress axis. The stress is plotted as a ratio to the applied stress σ_x. (The equal stress model was used for the transverse Young's modulus of a lamina.)

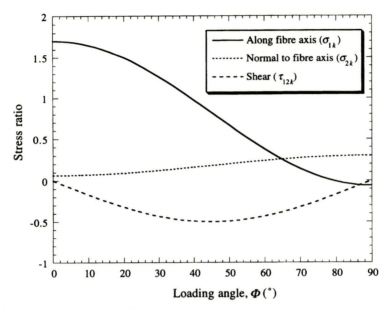

Fig. 5.12   Variation with loading angle Φ of the stresses σ_{1k}, σ_{2k} and τ_{12k} within the ply oriented at 0° to the stress axis, for an epoxy/50% glass fibre 0/90 (crossply) laminate. The stresses are plotted as a ratio to the applied stress. (The equal stress model was used for the transverse Young's modulus of a lamina.)

between the loading direction and the fibre direction in the 0° ply. This is shown for a single lamina and for two laminates. When more plies are present, so that the 0° ply being considered presents a smaller relative section, it bears a proportionately larger stress at $\Phi = 0$. This is because the 0° ply is much stiffer than those at other orientations. Note also that the ply can be put into compression when it is oriented at a large angle to the loading direction. This is the result of a Poisson contraction effect. It can also be seen in Fig. 5.12, which shows all the stresses present in the 0° ply for a crossply (0/90) laminate. The compressive stress parallel to the fibres (in the 1-direction), which has a magnitude of about 0.07 times the applied stress at $\Phi = 90°$, arises because the ply has a small natural Poisson contraction parallel to the fibres, but is being stressed by the larger natural contraction of the other ply (which is being loaded along its own fibre axis for $\Phi = 90°$).

The stresses in the two plies are illustrated schematically in Fig. 5.13 for $\Phi = 0°$. (Those shown for the 90° ply are the same as the data in

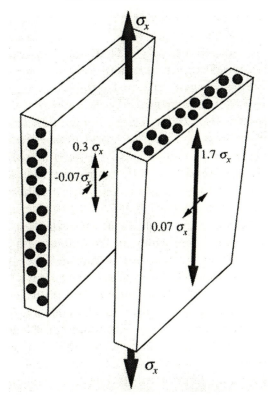

Fig. 5.13   Schematic representation of the stresses σ_{1k} and σ_{2k} within both plies of a 0/90 (crossply) epoxy/50% glass fibre laminate, oriented so that the fibres are normal and parallel to an applied tensile stress σ_x.

Fig. 5.12 for the 0° ply at $\Phi = 90°$.) It is clear from Fig. 5.11 that the compressive stresses generated in this way within individual plies can in some cases be surprisingly large. The details are sensitive to the Poisson's ratio of fibre and matrix, but – since (ceramic) fibres tend to have low values (relative to both polymers and metals) – the composite Poisson's ratio for contraction parallel to the fibre axis of a ply will always be less than that in a transverse direction and hence an effect of this type is quite general in composites based on these matrices.

5.4.3 Coupling stresses and symmetric laminates

Through-thickness (or 'coupling') stresses between the laminae are difficult to describe rigorously. However, they are important in practice and can lead to significant distortions of the laminate. The general nature of the distortions can be illustrated by two examples. Consider the simple crossply laminate illustrated in Fig. 5.14. If this is loaded in uniaxial tension, as in (a), the difference in natural Poisson contractions of the two plies will cause the laminate to distort in the manner shown, and out-of-plane stresses are needed to maintain the assembly flat. In addition to this, the transverse ply also exhibits a large through-thickness contraction as can be seen from the data in Fig. 4.10(a). A similar effect is shown in Fig. 5.14(b), for a crossply laminate which has been heated. Because the thermal expansion coefficients are different parallel and normal to the fibre axis (see §10.1.2), the laminate is deformed and becomes saddle-shaped. In this case it is assumed that $\alpha_2 > \alpha_1$, which is expected in view of the low thermal expansion coefficients exhibited by (ceramic) fibres. The stresses which arise within such a laminate on changing the temperature can also cause microstructural damage – see §10.1.4.

Distortions such as these can be reduced considerably if the arrangement of the plies is symmetric about the mid-plane of the laminate, i.e. if it has a mirror plane lying in the plane of the laminate. In *symmetric* laminates, the coupling forces largely cancel out and the laminate as a whole will not distort, although there are still local stresses across the interlaminar boundaries. In addition, the use of many thin laminae rather than a few thick ones minimises the distortions and leads to a reduction of the local interlaminar stresses. The classification of laminae according to whether or nor the stacking sequence is balanced and/or symmetric is illustrated in Fig. 5.15 with some examples.

There are many advantages in using *balanced symmetric* stacking sequences and this is common commercial practice. However, it should be noted that a laminate is often designed in the light of information about the expected in-service stress state. For example, with tubes to be subjected to internal or external pressure, unequal biaxial tension or compression will be imposed and ply angle sequences will be chosen with this in mind. The probable mode of failure, as well as the elastic deflections, may also need to be considered (see Chapter 8). Furthermore, the type and magnitude of permissible deflections and distortions will vary widely between different applications. It is therefore rather difficult to specify an optimum stacking sequence without detailed information about the performance requirements. This highlights the important concept of designing the material and the component simultaneously – a recurrent theme when working with composites.

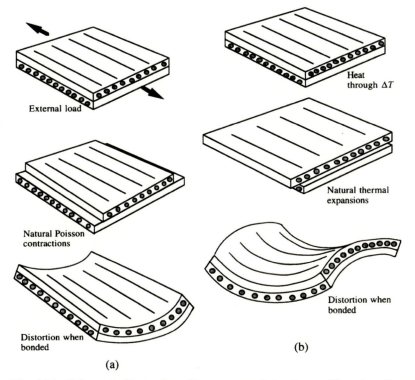

(a)

(b)

Fig. 5.14   Schematic illustration of how a crossply laminate will tend to distort as a result of through-thickness coupling stresses when subjected to (a) an external load and (b) a change in temperature.

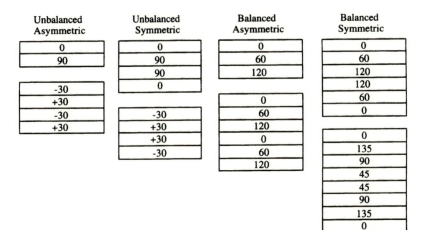

Unbalanced Asymmetric	Unbalanced Symmetric	Balanced Asymmetric	Balanced Symmetric
0	0	0	0
90	90	60	60
	90	120	120
	0		120
-30		0	60
+30		60	0
-30	-30	120	
+30	+30	0	0
	+30	60	135
	-30	120	90
			45
			45
			90
			135
			0

Fig. 5.15   Examples of stacking sequences which result in laminates classified according to whether they are balanced and/or symmetric.

References and further reading

Chou, T. W. (1989) Flexible composites, *J. Mater. Sci.*, **24** 761–83

Chou, T. W. and Kelly, A. (1980) Mechanical properties of composites, *Ann. Rev. Mater. Sci.*, **10** 229–59

Dieter, G. E. (1986) *Mechanical Metallurgy*. McGraw-Hill: New York

Gordon, J. E. (1978) *Structures*. Penguin: Harmondsworth, Middlesex

Jones, R. M. (1975) *Mechanics of Composite Materials*. McGraw-Hill: New York

Nye, J. F. (1985) *Physical Properties of Crystals – Their Representation by Tensors and Matrices*. Clarendon: Oxford

6

Stresses and strains in short-fibre composites

*The previous two chapters are concerned with the elastic behaviour of composites containing fibres which are, in effect, infinitely long. The preparation of composites containing short fibres (or equiaxed particles) allows scope for using a wider range of reinforcements and more versatile processing and forming routes (see Chapter 11). Thus, there is interest in understanding the distribution of stresses within such a composite, and the consequences of this for the stiffness and other mechanical properties. In this chapter, brief outlines are given of two analytical approaches to this problem. In the **shear lag model**, a cylindrical shape of reinforcements is assumed, and the stress fields in fibre and matrix are simplified so as to allow derivation of straightforward analytical expressions for the composite stiffness. The **Eshelby method**, on the other hand, is based on the assumption that the reinforcement has an ellipsoidal shape (which could range from a sphere to a cylinder or a plate). This allows derivation of an analytical solution which is more rigorous than that of the shear lag model, but with the penalty of greater mathematical complexity. In the treatment given here, attention is concentrated on the principle of the Eshelby approach; sources are suggested for readers needing more mathematical details.*

6.1 The shear lag model

The most widely used model describing the effect of loading an aligned short-fibre composite is the shear lag model, originally proposed by Cox (1952) and subsequently developed by others (Outwater 1956, Rosen 1960, Dow 1963), which centres on the transfer of tensile stress from matrix to fibre by means of interfacial shear stresses. The basis of the calculations is shown schematically in Fig. 6.1. In this diagram, reference

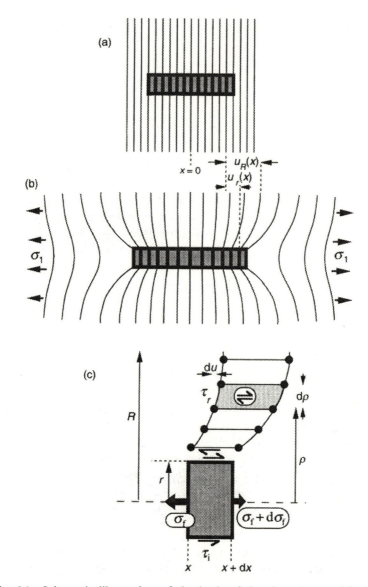

Fig. 6.1   Schematic illustration of the basis of the shear lag model, showing:
(a) unstressed system, (b) axial displacements *u* introduced on applying tension
parallel to the fibre and (c) variation with radial location of the shear stress
and strain in the matrix.

lines are drawn on a fibre and the surrounding matrix, which are initially straight and normal to the fibre axis. External loading is then applied (Fig. 6.1(b)) parallel to the fibre axis. The reference lines distort in the manner shown. Attention is concentrated on the shear distortions of the matrix close to the fibre, represented schematically in Fig. 6.1(c). The model is based on considering the shear stresses in the matrix and at the interface.

6.1.1 Stress and strain distributions

The radial variation of shear stress in the matrix, τ, (at a given axial distance x from the fibre mid-point) is obtained by equating the shear forces on neighbouring annuli (with radii r_1, r_2 of length dx (see Fig. 6.1(c))

$$2\pi r_1 \tau_1 \, dx = 2\pi r_2 \tau_2 \, dx$$
$$\therefore \frac{\tau_1}{\tau_2} = \frac{r_2}{r_1}$$

The shear stress τ in the matrix at any radius ρ is therefore related to that at the fibre/matrix interface (radius r), τ_i by

$$\tau = \tau_i \left(\frac{r}{\rho}\right)$$

The strain field around the fibre can be defined in terms of the **displacement** u of the matrix in the x-direction, relative to the position for no applied stress (see Fig. 6.1). The increment of this displacement, du, on moving out from the fibre axis by $d\rho$, is determined by the shear strain γ, and hence by the shear modulus G_m

$$\frac{du}{d\rho} = \gamma = \frac{\tau}{G_m} = \frac{\tau_i}{G_m}\left(\frac{r}{\rho}\right)$$

For any given value of x, the difference between the displacement of the matrix at a radius R and that of the interface is given by a simple integration

$$\int_{u_r}^{u_R} du = \frac{\tau_i r}{G_m} \int_r^R d\rho/\rho$$
$$(u_R - u_r) = \frac{\tau_i r}{G_m} \ln(R/r) \tag{6.1}$$

The matrix strain is assumed to be uniform remote from the immediate vicinity of the fibre. The radius R represents some far-field location where this condition becomes operative. In a composite containing an array of fibres, the appropriate value of (R/r) is related to the proximity of neighbouring fibres and hence to the fibre volume fraction, f. The exact relationship between (R/r) and f is dependent on the way the fibres are arranged: however, because (R/r) appears in a logarithmic term, the final result is relatively insensitive to the details of the fibre arrangement. An hexagonal array of fibres leads to

$$f = \frac{\pi r^2}{(2R)(R\sqrt{3})}$$

$$\left(\frac{R}{r}\right)^2 = \frac{\pi}{2f\sqrt{3}} \sim \frac{1}{f} \tag{6.2}$$

The build-up of tensile stress in the fibre σ_f is determined from the distribution of interfacial shear stress. Referring to Fig. 6.1(c), the basic force balance acting on an element of the fibre is

$$2\pi r \mathrm{d}x\tau_i = -\pi r^2 \mathrm{d}\sigma_f$$

$$\frac{\mathrm{d}\sigma_f}{\mathrm{d}x} = -\frac{2\tau_i}{r} \tag{6.3}$$

Now, the variation of τ_i with x is unknown a priori, but Eqn (6.1) can be used to relate it to displacements and hence to axial strains. It is assumed that there is no shear strain in the fibre and the interfacial adhesion is perfect (so that $u_r = u_f$, the displacement of the fibre surface). The following relationship is used for the shear modulus of the matrix

$$G_m = \frac{E_m}{2(1 + \nu_m)}$$

and substitution in Eqn (6.1) leads to the result

$$\frac{\mathrm{d}\sigma_f}{\mathrm{d}x} = -\frac{2E_m(u_R - u_f)}{(1 + \nu_m)r^2 \ln(1/f)} \tag{6.4}$$

The displacements themselves are unknown, but their differentials are related to identifiable strains. For the fibre

$$\frac{\mathrm{d}u_f}{\mathrm{d}x} = \varepsilon_f \left(= \frac{\sigma_f}{E_f} \right) \tag{6.5}$$

The corresponding expression for the matrix is less well defined. The differential of u_R will approximate to the far-field matrix strain, at least over most of the length of the fibre, and this, in turn, is close to the overall composite strain ε_1

$$\frac{\mathrm{d}u_R}{\mathrm{d}x} \approx \varepsilon_m \sim \varepsilon_1 \tag{6.6}$$

Although not rigorous, this represents a fairly good approximation; the far-field matrix strain is shown in Fig. 6.1(b) as being approximately uniform along (and beyond) the length of the fibre. The fibre strain (and stress) builds up with distance in from the ends of the fibre.

Finally, the stress distribution in the fibre can be determined. Differentiation of Eqn (6.4) and substitution leads to

$$\frac{\mathrm{d}^2\sigma_f}{\mathrm{d}x^2} = \frac{n^2}{r^2}(\sigma_f - E_f\varepsilon_1) \tag{6.7}$$

in which n is a dimensionless constant given by

$$n = \left[\frac{2E_m}{E_f(1 + \nu_m)\ln(1/f)}\right]^{1/2} \tag{6.8}$$

Eqn (6.7) is a standard second-order linear differential equation with the solution

$$\sigma_f = E_f\varepsilon_1 + B\sinh\left(\frac{nx}{r}\right) + D\cosh\left(\frac{nx}{r}\right) \tag{6.9}$$

On applying the boundary conditions

$$\sigma_f = 0 \quad \text{at} \quad x = \pm L$$

where L is the fibre half-length, and writing the fibre aspect ratio (L/r) as s, this gives the solution

$$\sigma_f = E_f\varepsilon_1[1 - \cosh(nx/r)\,\mathrm{sech}(ns)] \tag{6.10}$$

The variation of interfacial shear stress along the fibre length is derived, according to Eqn (6.3), by differentiating this equation, to give

$$\tau_i = \frac{n\varepsilon_1}{2} E_f \sinh\left(\frac{nx}{r}\right)\mathrm{sech}(ns) \tag{6.11}$$

6.1.2 The stress transfer length

Equations (6.10) and (6.11) allow predictions to be made about the stress distribution along the length of the fibre. An example is shown in Fig. 6.2(a) and (b). This shows the variations in fibre tensile stress and inter-facial shear stress along the length of a fibre in a composite of aligned

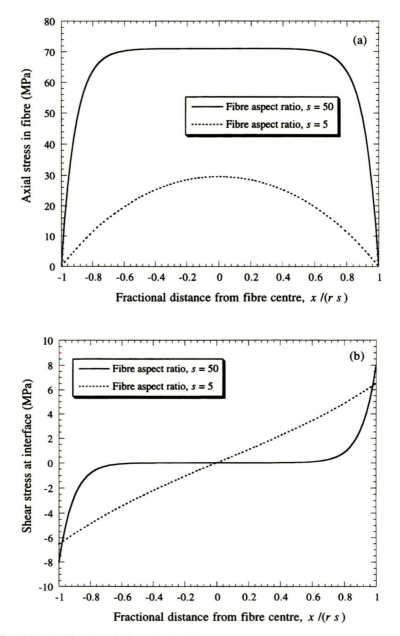

Fig. 6.2   Predicted variations in (a) fibre tensile stress and (b) interfacial shear stress along the length of a glass fibre in a polyester/30% glass fibre composite subject to an axial tensile strain of 10^{-3}, for two fibre aspect ratios.

glass fibres in a polyester resin matrix. The curves are for two different fibre aspect ratios and the composite has been subjected to a tensile strain parallel to the fibres of 0.1% ($\varepsilon_1 = 10^{-3}$). The tensile stress is zero at the fibre ends and a maximum in the centre. The interfacial shear stress is zero in the centre and a maximum at the ends. For the high aspect ratio case ($s = 50$), the fibre is long enough for the tensile stress to build up until the fibre has a strain equal to that of the matrix and the composite. This gives rise to the plateau region of the fibre stress curve and a region of zero interfacial shear stress. (With continuous aligned fibres, all of the composite is in this equal strain condition with respect to stress in the axial direction – see §4.1.) There are regions of the fibre near the ends which are less heavily stressed than this central plateau region, so that the average fibre stress is lower than in a long-fibre composite subjected to the same external load. The reinforcing efficiency decreases as the fibre length is reduced, since this increases the proportion of the total fibre length which is not fully loaded.

This behaviour leads to the concept of a **stress transfer length**, over which the strain in the fibre builds up to the plateau (matrix) value. For the case shown, this length is about 10 fibre diameters (i.e. about 40% of the distance from the end to the mid-point of the fibre for the $s = 50$ curve). Provided that the system remains fully elastic and there is no interfacial sliding, this value is dependent only on the elastic constants of fibre and matrix. (With a stiffer matrix, such as a metal, the stress transfer length will be shorter, as a result of higher interfacial shear stresses.)

For the low aspect ratio ($s = 5$) case shown in Fig. 6.2, the whole length of the fibre is only 5 fibre diameters, so that the stress in it does not build up to a plateau value. Such fibres are not providing very efficient reinforcement, because they carry much less stress than would longer fibres in the same system. A **stress transfer aspect ratio**, s_t, can be identified as that exhibited by fibres in which the peak (central) stress closely approaches the maximum possible (at which its strain is equal to the value being imposed on the composite). From Eqn (6.10), this will be the case when

$$\cosh(0)\,\mathrm{sech}(ns) \ll 1$$

Since $\cosh(0) = 1$, the requirement is to set $\mathrm{sech}(ns)$ ($= 1/\cosh(ns)$) to a suitably low value. Choosing 0.1 as being $\ll 1$, the condition becomes

$$\cosh(ns) \geq 10$$

$$\therefore s_t \approx \frac{3}{n} \tag{6.12}$$

The value of n, obtained from Eqn (6.8), becomes smaller as the fibre/matrix modulus ratio rises, and as the volume fraction of fibres decreases. In general, however, the value does not vary widely; it is typically about 0.1 for polymer matrix composites and 0.4 for metal matrix composites. The corresponding values of s_t are therefore of the order of 30 and 7, respectively. (For ceramic matrix composites, the value of n is normally close to unity and hence s_t is small. However, the introduction of ceramic fibres into a ceramic matrix is normally done for purposes other than that of stiffening the material, so that efficient load transfer is not a primary objective.)

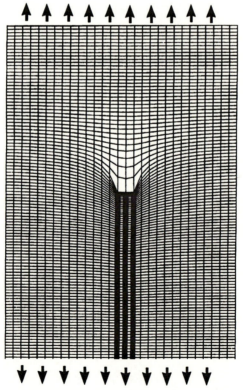

Fig. 6.3 An elastic deformation map obtained by a finite difference method (Termonia 1987), showing how an initially orthogonal grid around a fibre end becomes distorted on applying an axial tensile stress. (The fibre/matrix stiffness ratio = 40.)

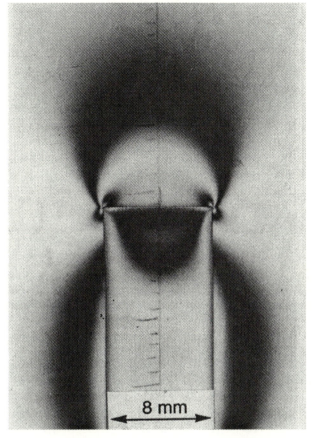

Fig. 6.4 Photoelastic fringe pattern (Withers *et al.* 1990) in and around a resin cylinder within a (more compliant) resin matrix subjected to axial stress. (The fringes represent contours of equal shear stress.)

The shear lag model has been shown to be qualitatively realistic. For example, using the finite difference method, Termonia (1987) illustrated how high matrix shear strains near the end of the fibre lead to a build-up of fibre tensile strain (see Fig. 6.3). This has also been confirmed by photoelastic observations. For example, the high-order fringes near the fibre ends in Fig. 6.4 correspond to regions of high shear strain. However, quantitative examination reveals discrepancies. The data in Fig. 6.5 were obtained (Galiotis *et al.* 1984) by measuring the local tensile strain at different points along a polydiacetylene fibre embedded in an epoxy matrix subjected to an external tensile load. While the general appearance

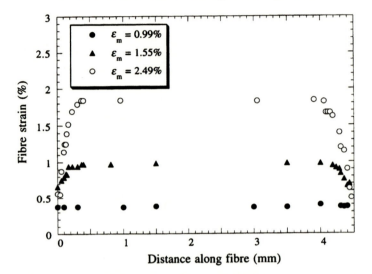

Fig. 6.5 Strains in a polydiacetylene fibre embedded in an epoxy matrix, measured by shifts in the Raman resonance spectrum (Galiotis *et al.* 1984), for three
values of the macroscopic strain applied to the matrix. The fibre aspect ratio was
about 200 and the fibre/matrix stiffness ratio was about 16.

of the curves agrees well with the shear lag model for the higher imposed
strains, there are discrepancies. Notable among these is that the fibre
stress does not fall to zero at the ends. This is primarily a result of the
transfer of tensile stress across the fibre ends, which is neglected in the
basic model and is relatively unimportant in composites with high fibre
aspect ratios. This is briefly examined in the next section.

6.1.3 Transfer of normal stress across fibre ends

Several attempts have been made (Fukuda and Chou 1981, Nardone and
Prewo 1986, Clyne 1989) to introduce corrections for the neglect of stress
transfer across the fibre ends. Any attempt to account for the effect, while
retaining the attractive simplicity of the shear lag approach, must involve
postulating an analytical expression for the fibre end stress σ_e. This must
be an arbitrary postulate, since there is no scope within the shear lag
framework for any rigorous description of stresses beyond the fibre
end. An example is provided by the suggestion (Clyne 1989) that σ_e be
set equal to the average of the peak fibre stress and the remote matrix
stress values predicted by the standard shear lag model

$$\sigma_e = \frac{\sigma_{f0} + \sigma_{m0}}{2}$$

in which σ_{f0} is given by substituting $x = 0$ in Eqn (6.10) and σ_{m0} is taken as $E_m \varepsilon_1$ (the average matrix stress). This leads to an expression for σ_e

$$\sigma_e = \frac{\varepsilon_1 [E_f(1 - \mathrm{sech}(ns)) + E_m]}{2} = \varepsilon_1 E'_m \qquad (6.13)$$

and hence, using the new boundary conditions $\sigma_f = \sigma_e$ at $x = \pm L$ to solve Eqn (6.9), a new expression is obtained for σ_f, analogous to Eqn (6.10)

$$\sigma_f = \varepsilon_1 [E_f - (E_f - E'_m) \cosh\left(\frac{nx}{r}\right) \mathrm{sech}(ns)] \qquad (6.14)$$

In Fig. 6.6, predictions from this equation are compared with those of the standard model (Eqn (6.10)) for (a) polymer- and (b) metal matrix composites. It can be seen that the predicted stresses in the fibre are significantly higher for the modified model, particularly near the fibre ends. Taking account of fibre end stress transfer naturally leads to the fibres carrying more load, particularly for short fibres. This results in an increase in the predicted stiffness of the composite (see next section).

6.1.4 Prediction of stiffness

The basic results of the shear lag treatment can be used to predict the elastic deformation of the composite. Consider a section of area A taken normal to the loading direction (in which all the fibres are aligned), as shown in Fig. 6.7. This section intersects individual fibres at random positions along their length. The applied load can be expressed in terms of the contributions from the two components

$$\sigma_1 A = f A \bar{\sigma}_f + (1 - f) A \bar{\sigma}_m$$
$$\therefore \sigma_1 = f \bar{\sigma}_f + (1 - f) \bar{\sigma}_m \qquad (6.15)$$

in which $\bar{\sigma}_f$ and $\bar{\sigma}_m$ are the volume-average stresses carried by fibre and matrix. This equation is often termed the '*Rule of Averages*'. The average fibre stress is evaluated from Eqn (6.10)

$$\bar{\sigma}_f = \frac{E_f \varepsilon_1}{L} \int_0^L \left[1 - \frac{\cosh(nx/r)}{\cosh(ns)}\right] \mathrm{d}x$$
$$\bar{\sigma}_f = E_f \varepsilon_1 \left(1 - \frac{\tanh(ns)}{ns}\right) \qquad (6.16)$$

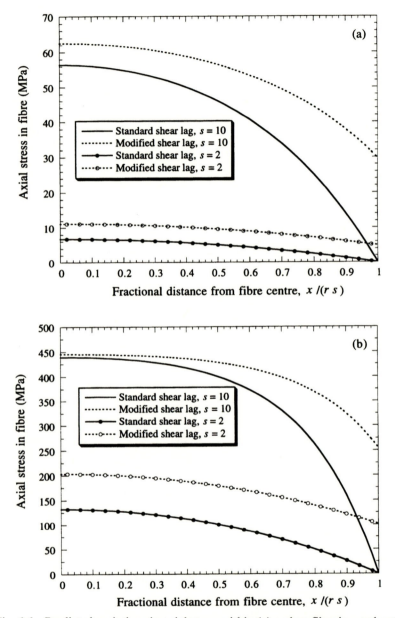

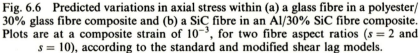

Fig. 6.6   Predicted variations in axial stress within (a) a glass fibre in a polyester/ 30% glass fibre composite and (b) a SiC fibre in an Al/30% SiC fibre composite. Plots are at a composite strain of 10^{-3}, for two fibre aspect ratios ($s = 2$ and $s = 10$), according to the standard and modified shear lag models.

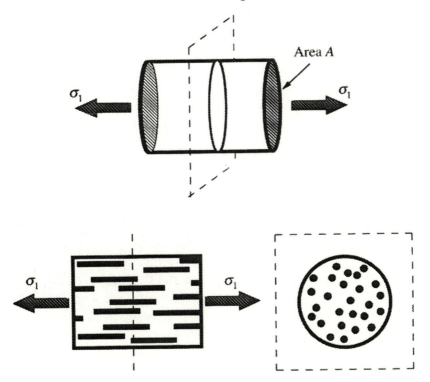

Fig. 6.7 Schematic illustration of a random section through a stressed short-fibre composite, showing how the section intersects individual fibres at various points along their length.

For the matrix, it is again conventional to resort to the assumption of a uniform tensile strain equal to that imposed on the composite

$$\bar{\sigma}_m \approx E_m \varepsilon_1 \qquad (6.17)$$

Combining Eqns (6.15)–(6.17) gives the stress–strain relationship for the composite

$$\sigma_1 = \varepsilon_1 \left[f E_f \left(1 - \frac{\tanh(ns)}{ns} \right) + (1-f)E_m \right] \qquad (6.18)$$

The same procedure for the modified model, taking account of fibre end stress transfer, leads to

$$\sigma_1 = \varepsilon_1 \left[f E_f \left(1 - \frac{(E_f - E'_m)\tanh(ns)}{E_f ns} \right) + (1-f)E_m \right] \qquad (6.19)$$

A linear increase in stress with increasing strain is predicted in both cases. The expressions in square brackets are the predicted Young's moduli for the two models.

These equations can be tested by making comparisons with predictions from the (more rigorous) Eshelby model (see §6.2). For example, Fig. 6.8 shows the variation in composite Young's modulus with fibre/matrix modulus ratio for two fibre aspect ratio values. It can be seen that the standard shear lag model is inaccurate for low fibre aspect ratios. The predictions of the standard model look particularly unreliable when the fibre/matrix modulus ratio is small. This suggests that the fibre end stress modification might be particularly useful for discontinuously reinforced metal matrix composites. This is confirmed by the data in Fig. 6.9, which compares predictions (McDanels 1985) from the three models with measured stiffnesses for particulate MMCs. The standard shear lag model is clearly quite unsuitable for application to such materials.

It may be noted from Eqns (6.18) and (6.19) that, as expected, the stiffness approaches the limiting (Rule of Mixtures) value as s becomes large enough for $\tanh(ns)/ns$ to become negligible ($\ll 1$). Since $\tanh(ns) \sim 1$ for $ns \gtrsim 3$, and assuming that 0.1 can be taken as $\ll 1$,

$$s_{RM} \approx \frac{10}{n} \qquad (6.20)$$

in which s_{RM} is the fibre aspect ratio needed for the composite modulus to approach its maximum (Rule of Mixtures) value. As noted earlier, values of n are typically around 0.1 for polymer composites and 0.4 for those with metallic matrices. This suggests values for s_{RM} of about 100 and 25 for PMCs and MMCs, respectively. These can be regarded as target (minimum) aspect ratios when the main objective is to maximise the load transfer and hence the stiffness.

6.1.5 Onset of inelastic behaviour

Several phenomena can occur which cause departure from ideal elastic behaviour. These include plastic deformation of the matrix, debonding and subsequent frictional sliding at the interface, formation of cavities or cracks in the matrix (particularly at fibre ends) and fracture of fibres. These effects change the stress distribution and hence affect the stress–strain curve. They are also related to the onset of failure and hence to the strength of the material. Detailed consideration of the factors involved is presented in Chapters 8 and 9, but it is appropriate here to examine a

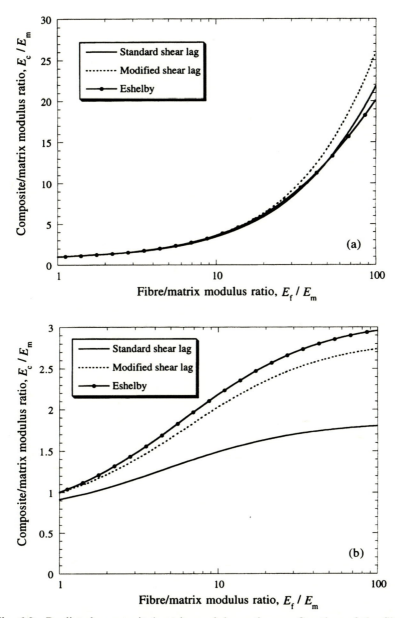

Fig. 6.8 Predicted composite/matrix modulus ratio as a function of the fibre/matrix modulus ratio, for composites with 30% reinforcement and fibre aspect ratios of (a) 30 and (b) 3. (Poisson's ratios of fibre and matrix were taken as 0.3 and 0.2 respectively.)

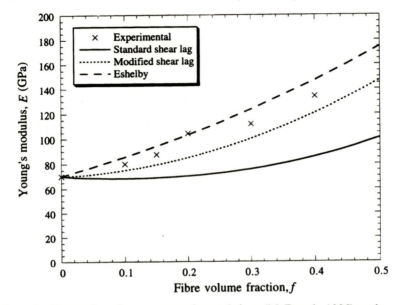

Fig. 6.9 Comparison between experimental data (McDanels 1985) and model predictions for the stiffness of Al/SiC (particulate) composites produced by extrusion. As the particles are not truly equiaxed, and tend to become aligned during processing, an aspect ratio of 2 was used in the predictions.

simple extension to the basic shear lag theory designed to predict the onset of departure from elastic behaviour.

The onset of matrix plasticity or interfacial sliding is expected to occur at the fibre ends, where the matrix shear stress is a maximum. A critical interfacial shear stress τ_{i^*} can be specified for these processes. Substitution of τ_{i^*} into Eqn (6.11), with $x = L$, gives the composite strain at the onset of such inelastic behaviour

$$\varepsilon_{1^*} = \frac{2\tau_{i^*}\coth(ns)}{nE_f} \qquad (6.21)$$

This can be converted to an applied stress using Eqn (6.18), leading to the expression

$$\sigma_{1^*} = \frac{2\tau_{i^*}}{nE_f}\left[f\,E_f + (1-f)E_m)\coth(ns) - \frac{f\,E_f}{ns}\right] \qquad (6.22)$$

This point does not correspond to a clearly identifiable composite yield stress, since yielding (or interfacial sliding) is only taking place in a small localised region. However, at this point the stress–strain curve will start to depart from a linear plot. As an illustration of the use of Eqn (6.21), in a typical glass fibre reinforced polymer composite, with $\tau_{i^*} = 20$ MPa, the composite strain at the onset of inelastic behaviour is about 0.6% for long fibres ($s \sim 30$) and about 0.3% for short fibres ($s \sim 5$).

The likelihood of fibre fracture taking place before matrix yielding or interfacial sliding can also be examined. The peak stress in the fibre at the onset of interfacial sliding or yielding is found from Eqn (6.10) by setting $x = 0$ and the composite strain to the value given by Eqn (6.21). This leads to

$$\sigma_{f0} = \frac{2\tau_{i^*}}{n} \left[\coth(ns) - \operatorname{cosech}(ns)\right] \tag{6.23}$$

Schematic plots of this relationship are shown in Fig. 6.10, which also gives an indication of the range of values expected for τ_{i^*} in metallic and polymeric matrices and for the fracture stress σ_{f^*} exhibited by ceramic fibres. (A distinction must be drawn between thermosetting and thermoplastic polymers: the former are brittle and tend to exhibit interfacial sliding and/or matrix microcracking, but not plastic yielding.) It is clear from this plot that, on increasing the load applied to either type of composite, yielding or sliding at the interface take place before fibres start to fracture.

As the composite strain is increased, yielding (or sliding) spreads along the length of the fibre, raising the tensile stress in the fibre as the interfacial shear stresses increase. Fracture of fibres may then become possible and a simple treatment can be used to explore the limit of this effect. If it is assumed that the interfacial shear stress becomes uniform at τ_{i^*} along the length of the fibre, then a **critical aspect ratio**, s_*, can be identified, below which the fibre cannot undergo any further fracture. This corresponds to the peak (central) fibre stress just attaining its ultimate strength σ_{f^*}, so that, by integrating Eqn (6.3) along the fibre half-length

$$s_* = \frac{\sigma_{f^*}}{2\tau_{i^*}} \tag{6.24}$$

It follows from this that a distribution of aspect ratios between s_* and $s_*/2$ is expected, if the composite is subjected to a large strain. The value of s_* ranges from over 100, for a polymer composite with poor interfacial bonding, to about 2–3 for a strong metallic matrix.

6.2 The Eshelby method

The method has its origins in some simple 'thought experiments' outlined by J. D. Eshelby in the 1950s (Eshelby 1957, Eshelby 1959). The power

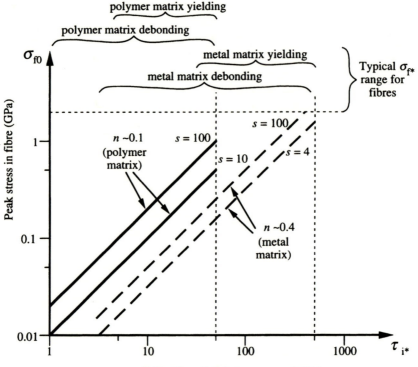

Fig. 6.10   Plots of the dependence of peak fibre stress σ_{f0} (at the onset of interfacial sliding or matrix yielding) on the critical shear stress for the onset of these
phenomena τ_{i^*} according to Eqn (6.23). Plots are shown for different aspect
ratios, with n values typical of polymer- and metal-based composites. Also indicated are typical value ranges for fracture of fibres and for matrix yielding and
interfacial debonding.

and versatility of the approach has only become widely appreciated quite
recently. The original study concerned the hypothetical situation of an
infinite matrix containing a stiff stress-free body. The body was removed
and then experienced a change of size and shape (a 'stress-free' strain).
The body was then subjected to surface forces so as to return it to its
original dimensions and replaced in the hole from whence it came. The
forces were then removed, allowing the body to adopt a new shape by
distorting the matrix. This, in effect, is what happens on heating a matrix
containing an inclusion, when the two have different coefficients of thermal expansion. (Other examples are provided by a martensitic transformation of the inclusion or a uniform plastic deformation of the matrix.)

The Eshelby method involves calculating the resultant stresses in the inclusion. These turn out to be uniform throughout the inclusion, provided it has an ellipsoidal shape. This is not as severe a limitation as it sounds, since many shapes approximate to ellipsoids. In particular, a short fibre can be represented by a prolate ellipsoid having the same aspect ratio. Derivation of expressions for stresses and strains in composites, and hence of their elastic constants, involves some manipulation of tensors. The treatment given here is highly abbreviated and readers interested in the details should consult other sources (Mura 1982, Taya 1988, Clyne and Withers 1993). However, the model can be used without a full understanding of the derivation.

6.2.1 A misfitting ellipsoid

Consider the sequence shown in Fig. 6.11(a). The initial mesh spacings of matrix and inclusion (fibre) represent unstrained material. The heavier lines of the inclusion denote its higher stiffness. For the case of differential thermal expansion on heating by ΔT, the imposed *misfit strain*, ε^{T^*}, is given by

$$\varepsilon^{T^*} = \Delta T(\alpha_f - \alpha_m) \qquad (6.25)$$

noting that thermal expansivity α and strain ε are both second-rank tensors. Since, in most cases, $\alpha_f < \alpha_m$, a decrease in temperature produces a change in inclusion shape, relative to the matrix, as shown in Fig. 6.11(a). After replacement of the inclusion in the cavity, it adopts a new size and shape, distorting the surrounding matrix as it does so. This new size and shape represents a *constrained strain*, ε^C, relative to the original state. A key point here is that, *for the special case of an ellipsoidal shape*, this strain is uniform in all parts of the inclusion. This must also be true of the stress, which can be written

$$\sigma_f = C_f(\varepsilon^C - \varepsilon^{T^*}) \qquad (6.26)$$

where C_f is the stiffness tensor of the inclusion (fibre) and the term in brackets represents the net strain relative to the 'stress-free' state after ε^T was imposed.

6.2.2 The equivalent homogeneous ellipsoid

The essence of the Eshelby method is to consider exactly the same operations being carried out with an ellipsoid having the *same elastic constants* *as the matrix*. This is shown in Fig. 6.11(b). The composite is now homo-geneous in terms of elastic properties. It is possible to generate exactly the same strain and stress state (in both inclusion and surrounding matrix) as arose with the real composite, provided that the imposed (stress-free) strain is the appropriate one. This is termed the *transformation strain*, ε^T. The schematic shown in Fig. 6.11(b) illustrates how choice of the appropriate value leads to the correct final stress state. The appropriate value of ε^T depends on, but differs from, the actual misfit strain ε^{T^*}.

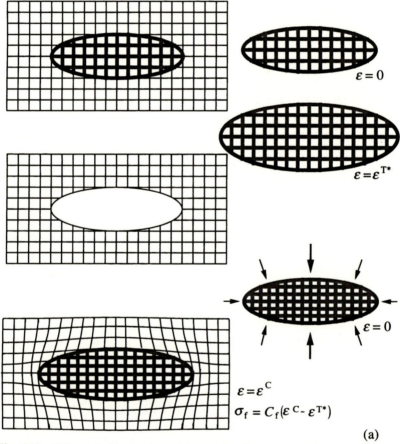

$$\varepsilon = 0$$

$$\varepsilon = \varepsilon^{T^*}$$

$$\varepsilon = 0$$

$$\varepsilon = \varepsilon^C$$

$$\sigma_f = C_f(\varepsilon^C - \varepsilon^{T^*})$$

(a)

Fig. 6.11 Schematic illustration of the strains during the Eshelby operations of removing an inclusion from the matrix, introducing a misfit strain ε^T and repla-cing it in the hole in the matrix. The operations are shown for (a) the actual inclusion and (b) an inclusion with the same elastic constants as the matrix – which must be subjected to a different strain ε^T in order to produce the same final stress state. (Grid spacing depicts elastic strain, while the stiffness is represented by the thickness of the grid lines.)

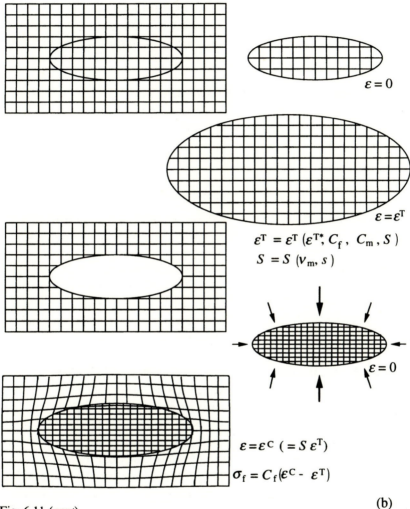

$$\varepsilon = 0$$

$$\varepsilon = \varepsilon^{\mathrm{T}}$$

$$\varepsilon^{\mathrm{T}} = \varepsilon^{\mathrm{T}} (\varepsilon^{\mathrm{T}*}, C_{\mathrm{f}}, \ C_{\mathrm{m}}, S \ )$$
$$S = S \ (v_{\mathrm{m}}, s \ )$$

$$\varepsilon = 0$$

$$\varepsilon = \varepsilon^{\mathrm{C}} \ (= S \ \varepsilon^{\mathrm{T}})$$
$$\sigma_{\mathrm{f}} = C_{\mathrm{f}}(\varepsilon^{\mathrm{C}} - \ \varepsilon^{\mathrm{T}})$$

(b)

Fig. 6.11 (*cont*)

(Other terms, such as the *eigenstrain*, ε^*, have been used by some authors to denote the appropriate transformation strain.)

The major contribution made by Eshelby was to establish that the constrained strain ε^{C} for this *equivalent homogeneous inclusion* is uniquely related to the transformation strain by the expression

$$\varepsilon^{\mathrm{C}} = S\varepsilon^{\mathrm{T}} \tag{6.27}$$

where S, the 'Eshelby tensor', is a simple function of the ellipsoid axis ratio (i.e. the aspect ratio of the fibre) and the Poisson's ratio of the matrix. (The symbol S is conventionally used for the Eshelby tensor; it is important to avoid confusion with S used to represent an elastic compliance.) Now, the stress in the homogeneous inclusion can be written

$$\sigma_f = C_m(\varepsilon^C - \varepsilon^T) \qquad (6.28)$$

This is set equal to the stress for the actual inclusion given by Eqn (6.26). Substitution for ε^C from Eqn (6.27) allows ε^T to be evaluated (for a given $\varepsilon^{T'}$, elastic constants and ellipsoid aspect ratio) and the stress in the inclusion to be calculated.

6.2.3 The background stress

The above treatment was developed for an infinite matrix and is only applicable to a 'dilute' composite, in which f is no more than, say, a few %. Extension to non-dilute composites is complicated by the fact that a given inclusion is affected by the matrix stress field resulting from neighbouring inclusions. This changes the stresses in both matrix and inclusion. The change may be regarded as a 'background stress', σ^b, superimposed on the stress state for a dilute case, although various other expressions (such as 'image stress') are also used to describe this stress. Several authors (Tanaka and Mori 1970, Brown and Clarke 1975, Mura 1982, Pedersen 1983) have contributed to the understanding and quantification of this effect, thereby increasing the usefulness of the Eshelby method enormously.

It can be shown that, if the matrix stresses are integrated over the whole of the volume around a misfitting inclusion, the result is zero. (This might have been expected from the symmetry of the situation, particularly on considering the example of a spherical inclusion.) The stress in the inclusion, however, is uniform and non-zero. This provides a means of evaluating the background stress in a non-dilute composite. For the equivalent homogeneous inclusion case, the background stress is expected to act uniformly throughout the composite, so that a force balance can be written

$$(1-f)\sigma^b + f(\sigma_f + \sigma^b) = 0 \qquad (6.29)$$

from which σ^b can be evaluated. For the case of the real composite, it is not clear how the inclusion experiences the background stress. The assumption is often made that the real and equivalent homogeneous inclusions are subject to the same strain disturbance as a result of being in a matrix containing other inclusions. This is the basis of the *'mean-field approximation'* and detailed arguments have been put forward to support the hypothesis. It should be quite accurate, at least for volume fractions up to a practical limit for short-fibre composites of around 40–50%. The term '*mean stress*' $\langle\sigma\rangle$, is now reserved for the volume-averaged stress in each component which contributes to this force balance

$$(1 - f)\langle\sigma\rangle_m + f\langle\sigma\rangle_f = 0 \tag{6.30}$$

The mean stresses are indicative of load transfer: the inclusion carries a mean stress of the opposite sign (and different magnitude) to that in the matrix.

From this point, some mathematical manipulations allow the mean strains of the two components, and hence of the composite, to be evaluated. For the case of a differential thermal expansion, the overall average composite strain, (equal to the sum of this mean strain and the strain from the normal expansion of the matrix) then determines the composite expansivity. (Thermal expansion is examined in §10.1.)

6.2.4 Composite stiffness

The analysis can be adapted to treat the case where the composite is subject to an applied stress, σ^A, and hence to predict the stiffness. The appropriate Eshelby operations are shown in Fig. 6.12(a) and (b). The misfit now effectively arises from the difference between the shapes which would be adopted by the inclusion (fibre) and by the cavity in the matrix, if the two constituents were to be independently subjected to the applied stress. For the equivalent homogeneous composite, the appropriate transformation strain ε^T is now dependent on σ^A, rather than on a pre-defined $\varepsilon^{T'}$. Furthermore, the resultant strain is now the sum of the constrained strain ε^C and a strain arising directly from the applied stress ε^A ($= C_m^{-1}\sigma^A$). Similarly, the total stress in the inclusion is made up of σ^A and the contribution introduced by the *load transfer* between the constituents, represented by σ_f (as in the misfit strain treatment). The same technique of equating the stresses in the actual and equivalent homogeneous inclusions

$$\begin{aligned}\sigma_f + \sigma^A &= C_f(\varepsilon^C + \varepsilon^A) \\ &= C_m(\varepsilon^C - \varepsilon^T + \varepsilon^A)\end{aligned} \tag{6.31}$$

and substitution of $S\varepsilon^T$ for ε^C, allows evaluation of ε^T as before.

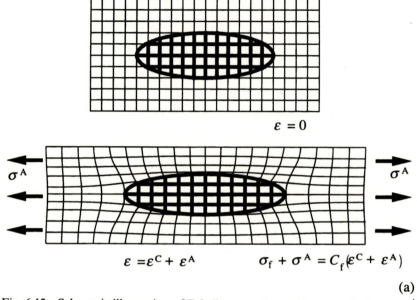

$$\varepsilon = 0$$

$$\varepsilon = \varepsilon^C + \varepsilon^A \qquad \sigma_f + \sigma^A = C_f(\varepsilon^C + \varepsilon^A)$$

(a)

Fig. 6.12   Schematic illustration of Eshelby operations with an applied stress σ^A for (a) actual and (b) equivalent homogeneous composites.

For non-dilute composites under applied stress, the force balance of Eqn (6.30) still holds, with the proviso that the mean stress $\langle \sigma \rangle$ in each component is not now the actual volume-averaged stress $\bar{\sigma}$, but is rather a measure of the difference between this and σ^A

$$\overline{\sigma_m} = \langle \sigma \rangle_m + \sigma^A \qquad (6.32)$$

$$\overline{\sigma_f} = \langle \sigma \rangle_f + \sigma^A \qquad (6.33)$$

The term '*mean stress*, $\langle \sigma \rangle$', therefore has a rather special meaning in Eshelby analysis, closely related to the nature of the load-sharing between the two components: the volume-averaged stress in the inclusions is greater than that over the composite as a whole by $\langle \sigma \rangle_f$, while the volume-averaged matrix stress is less than the overall average by $\langle \sigma \rangle_m$ (i.e. $\langle \sigma \rangle_m$ is negative).

After some mathematical manipulation, an expression can be derived for the stiffness tensor of the composite

$$C_c[[C_m^{-1} - f\{(C_f - C_m)[S - f(S - I)] + C_m\}^{-1}(C_f - C_m)C_m^{-1}]]^{-1}$$

(6.34)

in which I is the identity tensor.

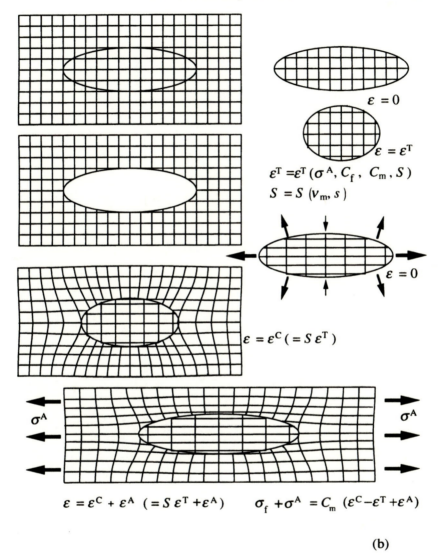

$$\varepsilon = 0$$

$$\varepsilon = \varepsilon^T$$

$$\varepsilon^T = \varepsilon^T (\sigma^A, C_f, C_m, S)$$

$$S = S (v_m, s)$$

$$\varepsilon = 0$$

$$\varepsilon = \varepsilon^C (= S \varepsilon^T)$$

$$\sigma^A \qquad\qquad \sigma^A$$

$$\varepsilon = \varepsilon^C + \varepsilon^A \ (= S \varepsilon^T + \varepsilon^A) \qquad \sigma_f + \sigma^A = C_m (\varepsilon^C - \varepsilon^T + \varepsilon^A)$$

(b)

Fig. 6.12 *(cont)*

The engineering constants of the composite can be derived from the stiffness tensor, which is best evaluated with a computer program. Typical results are shown in Fig. 6.13, which gives axial and transverse stiffness predictions for a polymeric composite. Fig. 6.13(a) confirms that, in the practicable volume fraction range up to about 40–50%, fibres with fairly high aspect ratios are needed in order to effect substantial

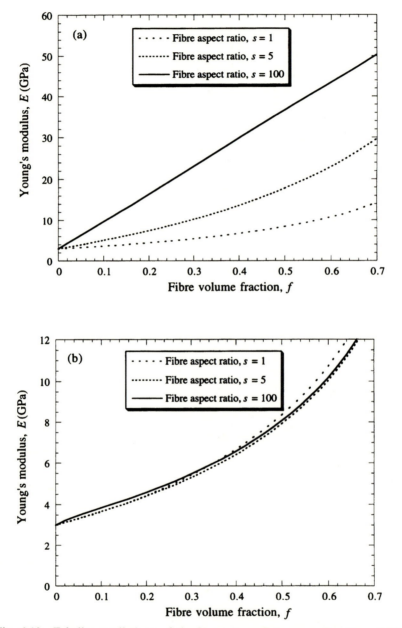

Fig. 6.13 Eshelby predictions of the Young's modulus as a function of fibre volume fraction for glass fibres with aspect ratios of 1, 5 and 100, in an epoxy matrix for (a) axial and (b) transverse loading.

improvements in the stiffness. The transverse stiffness predictions in Fig. 6.13(b), however, show clearly that the aspect ratio has very little effect on the transverse stiffness. This is the case for all composites.

References and further reading

Shear lag treatments

Clyne, T. W. (1989) A simple development of the shear lag theory appropriate for composites with a relatively small modulus mismatch, *Mat. Sci. & Eng.* **A122** 183–92

Cox, H. L. (1952) The elasticity and strength of paper and other fibrous materials, *Brit. J. Appl. Phys.*, **3** 72–9

Dow, N. F. (1963) *Study of Stresses near Discontinuity in a Filament-reinforced Composite Metal*, GE Co., Missile and Space div., Report No. R635D61

Fukuda, H. and Chou, T. W. (1981) An advanced shear lag model applicable to discontinuous fiber composites, *J. Comp. Mat.*, **15** 79–91

Nardone, V. C. and Prewo, K. M. (1986) On the strength of discontinuous silicon carbide-reinforced aluminium composites, *Scripta Met.*, **20** 43–8

Outwater, J. O. (1956) The mechanics of plastics reinforced in tension, *Modern Plastics*, **33** 56–65

Rosen, B. W. (1960) Mechanics of fibre strengthening, in *Fibre Composite Materials*. B. W. Rosen (ed.) ASM: Metals Park, Ohio, chapter 3

Eshelby modelling

Brown, L. M. and Clarke, D. R. (1975) Work hardening due to internal stresses in composite materials, *Acta Metall.*, **23** 821–30

Clyne, T. W. and Withers, P. J. (1993) *An Introduction to Metal Matrix Composites*. Cambridge University Press: Cambridge

Eshelby, J. D. (1957) The determination of the elastic field of an ellipsoidal inclusion, and related problems, *Proc. Roy. Soc.*, **A241** 376–96

Eshelby, J. D. (1959) The elastic field outside an ellipsoidal inclusion, *Proc. Roy. Soc.*, **A252** 561–9

Mura, T. (1982) *Micromechanics of Defects in Solids*. Martinus Nijhoff

Pedersen, O. B. (1983) Thermoelasticity and plasticity of composites – 1. Mean field theory, *Acta Metall.*, **31** 1795–808

Tanaka, K. and Mori, T. (1970) The hardening of crystals by non-deforming particles and fibres, *Acta Metall.*, **18** 931–9

Taya, M. (1988) Modelling of physical properties of metallic and ceramic composites; generalised Eshelby model, in *Mechanical and Physical Behaviour of Metallic and Ceramic Composites*. S. I. Andersen *et al.* (eds.) Risø Nat. Lab.: Roskilde, Denmark pp. 201–31

Withers, P. J., Smith, A. N., Clyne, T. W. and Stobbs, W. M. (1990) A photographic examination of the validity of the Eshelby approach to the modelling of MMCs, in *Fundamental Relationships between Microstructural and Mechanical Properties of Metal Matrix Composites*. M. N. Gungor and P. K. Liaw (eds.) TMS pp. 225–40

General

Carrara, A. S. and McGarry, F. J. (1968) Matrix and interfacial stresses in a discontinuous fibre composite model, *J. Comp. Mat.*, **2** 222–43

Galiotis, C., Young, R. J., Yeung, P. H. J. and Batchelder, D. N. (1984) The study of model polydiacetylene/epoxy composites. Part I: The axial strain in the fibre, *J. Mat. Sci.*, **19** 3640–8

McDanels, D. L. (1985) Analysis of stress–strain, fracture, and ductility of aluminium matrix composites containing discontinuous silicon carbide reinforcement, *Metall. Trans.*, **16A** 1105–15

Termonia, Y. (1987) Theoretical study of the stress transfer in single fibre composites, *J. Mat. Sci.*, **22** 504–8

7

The interface region

The preceding three chapters have dealt with the elastic behaviour of composites. Among the assumptions made in most of these treatments is that the interfacial bond is 'perfect'. This means that there is no debonding, cracking or sliding – in fact, no elastic or inelastic processes of any description. In practice, many important phenomena may take place at the interface, depending on its structure and the stresses generated there. These processes tend to promote plastic deformation of the matrix and can also influence the onset and nature of failure. Before treating the strength and fracture behaviour of composites (Chapters 8 and 9), it is necessary to consider the interface and examine how its response can be characterised and influenced. In the present chapter, the meaning and measurement of bond strength are described. This is followed by an outline of the formation of interfacial bonds in various systems and a summary of the techniques used to influence the bonding characteristics.

7.1 Bonding mechanisms

7.1.1 Adsorption and wetting

If the surfaces of two bodies spontaneously come into intimate (atomic scale) contact when they are brought close to each other (commonly with one of the bodies in liquid form), then 'wetting' is said to have taken place. Adhesion is primarily caused by van der Waals forces, although other types of bonding may reinforce these. The occurrence of wetting can be treated using simple thermodynamics, but in practice there may be chemical changes taking place which are time-dependent. Fig. 7.1 illustrates solid/solid and solid/liquid interfaces. The solid/solid contact area

133

The interface region

(a)

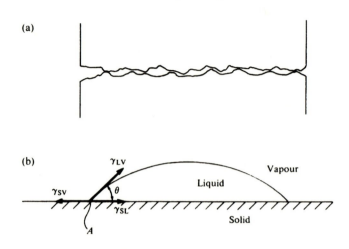

(b)

Fig. 7.1   (a) Isolated contact points leading to weak adhesion between two rigid rough surfaces. (b) Contact angle θ and surface energies γ for a liquid drop on a solid surface.

is limited to those regions where asperities touch (Fig. 7.1(a)) and the effective bond strength is very low unless extensive deformation is promoted to remove the asperities. Surface contamination can also restrict the area of effective contact. For the liquid/solid case, intimate contact can be established providing the liquid is not too viscous and a thermodynamic driving force exists. This is commonly expressed in terms of surface energies, γ, so that the **work of adhesion**, W_a, is a simple net sum, often termed the **Dupré equation**

$$W_a = \gamma_{SV} + \gamma_{LV} - \gamma_{SL} \tag{7.1}$$

The subscripts S, L and V refer to solid, liquid and vapour, respectively. The vapour phase is commonly air. According to this equation, wetting is strongly favoured if the surface energies of the two constituents are large and their interfacial surface energy is small. In practice, however, a large value of the liquid surface energy inhibits the spreading of a liquid droplet. The equilibrium wetting or **contact angle** θ is dictated by the **Young equation**, obtained by a balance of horizontal forces, Fig. 7.1(b),

$$\gamma_{SV} = \gamma_{SL} + \gamma_{LV} \cos \theta \tag{7.2}$$

It follows that complete wetting ($\theta = 0°$) occurs if the surface energy of the solid is equal to or greater than the sum of the liquid surface energy and the interface surface energy. Interface surface energies are difficult to obtain (and may be influenced by chemical reactions), but they are frequently smaller than the values for the phases being exposed to air. The surface energies of fibres and (liquid) matrices are generally known and systems where the former greatly exceeds the latter are likely to wet very easily. For example, glass ($\gamma_{SV} = 560$ mJ m^{-2}) and graphite ($\gamma_{SV} = 70$ mJ m^{-2}) are readily wetted by polyester ($\gamma_{LV} = 35$ mJ m^{-2}) and epoxy ($\gamma_{LV} = 43$ mJ m^{-2}) resins, but polyethylene fibres ($\gamma_{SV} = 31$ mJ m^{-2}) are not. Lack of wetting is a problem for certain metal matrix composites and coatings on the fibres are used to improve this (see §7.3.1).

7.1.2 Interdiffusion and chemical reaction

Various types of diffusional process which promote adhesion can take place at the interface. For example, Fig. 7.2(a) shows the diffusion of free chain ends at the interface between two polymers, which leads to chain entanglements and a rise in the adhesive strength. This effect is employed in some coupling agents used on fibres in thermoplastic matrices (see §7.3.1). Interdiffusion can also take place in non-polymeric systems, particularly if it is accompanied by a chemical reaction. The adhesive strength is dependent on the nature of the resultant interatomic bonds (and also on the stresses generated by the reaction – see below).

Various types of chemical reaction may occur at the interface, either deliberately promoted or inadvertent. These can be represented, as in Fig. 7.2(d), by new A–B bonds being formed as a result of interfacial chemical reactions. These bonds may be covalent, ionic, metallic, etc., and in many cases are very strong. There are many examples of the interfacial bond strength being raised by localised chemical reactions, but it is often observed that a progressive reaction occurs which results in the formation of a brittle reaction product.

Carbon fibres are prone to surface reactions with organic groups; the details depend on the manufacturing methods (e.g. see Scola 1974). An important feature is the angle at which the basal planes meet the free surface (see §2.1.1), as many reactions take place preferentially at the edges of these planes. For example, the high-modulus PAN-based fibres have a thick skin with basal planes predominantly parallel to the surface; reaction takes place less readily than in carbon fibres with basal planes normal to the surface and the fibres are prone to cohesive failure as a result of weak inter-plane bonding. Heat treatments prior to composite

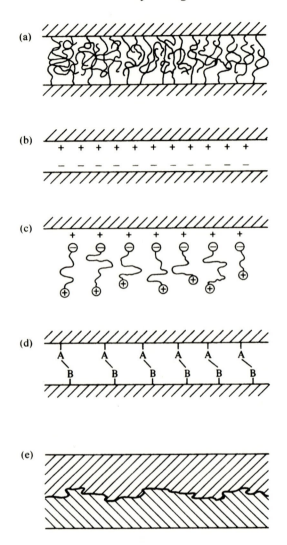

Fig. 7.2 Interfacial bonds formed by (a) molecular entanglement following interdiffusion, (b) electrostatic attraction, (c) cationic groups at the end of molecules attracted to an anionic surface, resulting in polymer orientation at the surface, (d) chemical reaction and (e) mechanical keying.

fabrication raise the bond strength as a result of oxidation of the fibres and removal of the surface layers.

There are several examples of bond strength enhancement by localised chemical reaction in MMCs and CMCs. The 'Saffil^{TM}' δ-alumina fibre contains a few % of silica, concentrated in the free surface and grain boundary regions. During exposure of this fibre to molten metals containing strong reducing agents, such as magnesium, the surface is attacked to the depth of silica enrichment (a few nanometres) – e.g. see Cappleman *et al.* (1985). This has been correlated with a considerable enhancement of bond strength compared with cases where the matrix is free of magnesium (Clegg *et al.* 1988), presumably as a result of covalent and/or ionic bonds extending across the interface. A similar effect has been observed during the early stages of the reaction in Ti/SiC. However, in this case the reaction is a progressive one which tends to result in a relatively thick layer of brittle reaction product (a mixture of Ti_5Si_3 and TiC – see Martineau *et al.* 1984).

7.1.3 Electrostatic attraction

If the surfaces carry net electrical charges of opposite sign, as illustrated in Fig. 7.2(b), then a sustained adhesive force may result. This effect is utilised in certain fibre treatments, as in the deposition of coupling agents on glass fibres (see §7.3.1). The surface may exhibit anionic or cationic properties, depending on the oxide in the glass and the pH of the aqueous solution used to apply the coupling agents. Thus, if ionic functional silanes are used, it is expected that the cationic functional groups will be attracted to an anionic surface and vice versa (Fig. 7.2(c)). Electrostatic forces are unlikely to constitute the major adhesive bond in a composite and they can readily be reduced, for example by discharging in the presence of a strongly polar solvent, such as water.

7.1.4 Mechanical keying

There may be a contribution to the strength of the interface from the surface roughness of the fibres if good wetting has occurred, as illustrated in Fig. 7.2(e). The effects are much more significant under shear loading than for decohesion as a result of tensile stresses. Some improved resistance to tensile failure results if re-entrant angles are present and there is an increase in strength under all types of loading as a consequence of the increased area of contact.

7.1.5 Residual stresses

The nature of the interfacial contact is strongly influenced by the presence of residual stresses. These arise from a number of sources, such as plastic deformation of the matrix and phase transformations involving volume changes. There are also volume changes associated with the curing of thermosetting resins. One of the most important sources of residual stress is the thermal contraction which occurs during post-fabrication cooling. Since, for most composite systems, the fibre has a smaller thermal expansivity (thermal expansion coefficient) than the matrix, the resultant stresses are compressive in the fibre and tensile in the matrix. This arises because the matrix contracts onto the fibre and compresses it. The nature of the stress field is illustrated by Fig. 7.3. This shows the principal stresses (in axial, hoop and radial directions) for a long glass fibre surrounded by a tube of polyester resin, after cooling through 100 °C. (The stress field in a multi-fibre system, which cannot be predicted from the analytical model used to generate Fig. 7.3, is very similar – except near the outer surface.)

In polymer matrix composites, some of the stresses generated by differential contraction are relaxed by viscoelastic flow or creep in the matrix. In a metal matrix composite, the temperature changes during cooling are greater, and the matrices are usually quite resistant to creep and plastic flow, so that the residual stresses are substantially higher. The normal stresses across the interface (radial stresses) are compressive. This is particularly relevant to the interfacial bonding, since this compression ensures that fibre and matrix are kept in close contact and increases the resistance to debonding and sliding. A relatively high interfacial shear strength results when the residual thermal stresses are large. However, the nature of the stresses is different if the fibre volume fraction is so high that the matrix becomes broken up into isolated regions, surrounded on all sides by fibres. In this case, the matrix tends to contract away from the fibres, leading to a tensile normal stress across the interface.

7.2 Experimental measurement of bond strength

The nature of the interfacial bonding affects the elastic and fracture properties of the composite in a number of ways. Single-fibre experiments are used to obtain quantitative information about the bond strength. Most measurements of bond strength involve shear debonding and sliding (often using simple variants of the shear lag theory to interpret the

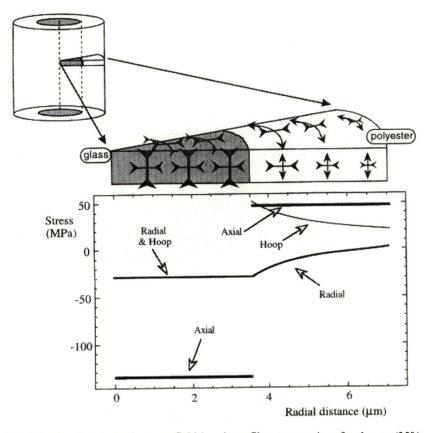

Fig. 7.3   Predicted elastic stress field in a long-fibre composite of polyester/35%
glass fibre, according to the analytical coaxial cylinder model of Mikata and Taya
(1985), showing the effect of cooling through a temperature interval of 100 K.
Property data used in producing these plots are given in Tables 2.2 and 2.5.

data), with little or no attempt to change the normal stress across the
interface. This is because of the difficulty in applying such stresses to a
cylindrical interface in a controlled manner. It might be argued that an
interface exhibiting a high shear debonding stress would also be expected
to offer strong resistance to a normal tensile stress. That this is not
necessarily true can be seen from the fact that pronounced interfacial
roughness is expected to raise the former (in shearing mode) while having
little effect on the latter (in opening mode). In relating data from interfacial tests to the macroscopic behaviour of the composite, it should
therefore be borne in mind that the two cases may involve different

interfacial stress states. Furthermore, some tests use artificial single-fibre 'composites'. The interfaces in these specimens may differ from those in the corresponding real materials, because different manufacturing techniques are used and there are different degrees of constraint in the absence of neighbouring fibres. Finally, while interpretation of test data is commonly carried out in terms of critical stress levels, an energy-based (fracture mechanics) analysis (Kendall 1985, Evans *et al.* 1990, Evans and Dalgleish 1993) may be more useful for some purposes – see Chapter 9.

7.2.1 Single-fibre pull-out test

This has been extensively applied to polymer composites. A single fibre, half embedded within a matrix, is extracted under a tensile load. The load–displacement data can be interpreted using an adaptation of the shear lag theory (Lawrence 1972, Chua and Piggott 1985). A schematic illustration of the axial distributions of normal stress in the fibre and shear stress at the interface is given in Fig. 7.4. These distributions are shown at three stages of pull-out, corresponding to elastic loading up to debonding, propagation of the debonding front and subsequent pull-out by frictional sliding. Basic assumptions of the shear lag model (see §6.1.1), such as no shear strain in the fibre and no transfer of normal stress across the fibre end, are retained in most treatments of this problem.

It is conventional to assume that the peak in the load–displacement plot corresponds to the debonding event, occurring at an applied stress σ_0, see Fig. 7.4. The treatment parallels that in §6.1, but with the ratio of the radius of the matrix to that of the fibre, R/r, retained rather than written in terms of the fibre volume fraction. The analogous equation to Eqn (6.4) is therefore

$$\frac{\mathrm{d}\sigma_f}{\mathrm{d}x} = \frac{E_m(u_R - u_r)}{(1 + \nu_m)r^2 \ln(R/r)} \tag{7.3}$$

The displacement conditions replacing Eqns (6.5) and (6.6) are now

$$\frac{\mathrm{d}u_r}{\mathrm{d}x} = \epsilon_f \tag{7.4}$$

$$\frac{\mathrm{d}u_R}{\mathrm{d}x} = 0 \tag{7.5}$$

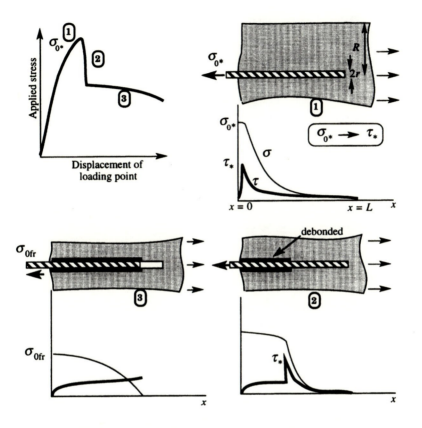

Fig. 7.4   Schematic stress distributions and load–displacement plot during the single-fibre pull-out test. The applied load generates an interfacial shear stress, which has a peak near to the front surface. At some critical applied load σ_0. this shear stress causes the interface to debond. Debonding then spreads along the interface and subsequent interfacial motion is by frictional sliding.

which corresponds to the interface being perfectly bonded (up until the debonding point) and the matrix remote from the interface being unstrained. The second-order linear differential equation governing the variation of σ_f along the length of the fibre is now slightly simpler than Eqn (6.7)

$$\frac{\mathrm{d}^2\sigma_f}{\mathrm{d}x^2} = \frac{n^2}{r^2}\sigma_f \qquad (7.6)$$

where the dimensionless constant is given by

$$n = \left[\frac{E_m}{E_f(1 + \nu_m)\ln(R/r)}\right]^{1/2} \tag{7.7}$$

Applying the boundary conditions $\sigma_f(0) = \sigma_0$ (where the fibre emerges from the matrix) and $\sigma_f(L) = 0$ (no stress transfer across the embedded fibre end) leads to

$$\sigma_f = \sigma_0 \left\{\frac{\sinh[n(L-x)/r]}{\sinh(nL/r)}\right\} \tag{7.8}$$

The interfacial shear stress, according to the basic equation of the shear lag model (Eqn (6.3)), then becomes

$$\tau = -\frac{r}{2}\frac{d\sigma}{dx} = \frac{n\sigma_0}{2}\cosh\left[\frac{n(L-x)}{r}\right]\text{cosech}\left(\frac{nL}{r}\right) \tag{7.9}$$

and on applying this at $x = 0$, the debonding shear stress, τ_* is deduced from the peak fibre stress (σ_{0^*})

$$\tau_* = \frac{n\sigma_{0^*}\coth(nL/r)}{2} \tag{7.10}$$

Some variants of the basic model have been published. For example, Hsueh (1990) has incorporated the possibility of stress transfer across the fibre end and ensured that the load carried by the free fibre is balanced by that in the composite. This model leads to more complex equations, but the predictions are similar. In particular, the ratio of τ_* to σ_{0^*} is usually very close to that for the basic shear lag treatment. Neither of these models takes account of the fact that the shear stress in the matrix should fall to zero at the free surface from which the fibre emerges. It has been shown (Grande *et al.* 1988) that the build up to the peak value typically takes place over a distance equal to about one-quarter of a fibre diameter. This point is considered further in the next section. Models have also been devised for the frictional sliding part of the pull-out curve, but these are more complex than for the debonding event.

A large data base now exists for pull-out testing of polymer matrix composites (e.g. see Favre 1989). Values for the debonding shear stress generally fall in the range 5–100 MPa. There are, however, practical difficulties in specimen preparation and handling, particularly when applying the procedure to composites with a relatively stiff matrix. In general, it is more convenient to use the push-out test described below.

7.2.2 Single-fibre push-out and push-down tests

These tests are much easier to apply to pieces of actual composite than is the pull-out test. The basis of the method is the application of a compressive axial load to the top surface of an embedded fibre until debonding occurs. In the *push-out* test, the specimen is in the form of a thin slice, with the fibre axis normal to the plane of the slice. The fibre becomes displaced so that it protrudes from the bottom of the specimen. The scanning electron microscopy (SEM) micrographs in Fig. 7.5 show SiC monofilaments that have been pushed out in this way. The test is easily applied to such large diameter fibres, but for fine (and strongly bonded) fibres there can be difficulties in preparing and handling thin section specimens. In the *push-down* (or indentation) test, the specimen is in bulk form and debonding is followed by the fibre frictionally sliding downwards over a certain distance, usually leaving a permanent displacement between the top of the fibre and the top of the matrix when the applied load is removed. The test was developed and analysed by Marshall (1985) and Marshall and Oliver (1987).

Schematic illustrations of the stress distributions and load–displacement curve are shown in Fig. 7.6 for the push-out test. Note that the Poisson effect now raises, rather than lowers, the frictional sliding stress. Analyses have been based on a simple shear lag approach. For example, Hsueh (1990) has presented models for both the debonding and frictional sliding behaviour. More recently, finite element method (FEM) techniques have been applied, in order to examine the stress field in more detail (Kerans and Parthasarathy 1991, Kallas *et al.* 1992, Kalton *et al.* 1994). An important conclusion from this work is that the shear lag models significantly overestimate the peak interfacial shear stress, for a given applied tensile load. This is illustrated by the data in Fig. 7.7, which compares FEM and shear lag predictions of the shear stress distribution with experimental data from a 'macroscopic' push-out set-up, constructed from two photoelastic resins. The peak of the shear lag curve is apparently too high by a factor of three in this case. An error of similar magnitude is predicted for the pull-out test.

In fact, since the FEM work has indicated that the distribution of interfacial shear stress tends to be more uniform than the shear lag

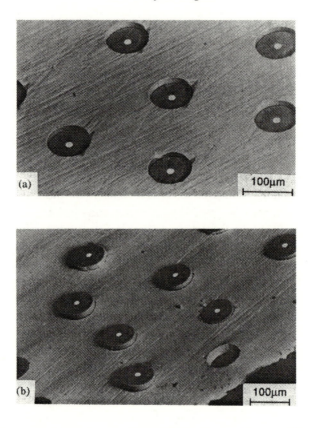

Fig. 7.5   Scanning electron micrographs (Watson and Clyne 1992) of a wedge-shaped Ti–6Al–4V/30% SiC monofilament specimen after single-fibre push-out testing, showing (a) the top surface and (b) the underside near the thin end of the wedge.

models predict, it may be acceptable in many cases to take it as constant. A simple force balance can then be used to obtain τ_* from the stress applied to the fibre σ_{0^*}.

$$\sigma_{0^*} \cdot \pi r^2 = \tau_* L 2\pi r$$
$$\therefore \tau_* = \frac{\sigma_{0^*}}{4s} \tag{7.11}$$

where the fibre aspect ratio, s, is given by its length, L, divided by its diameter. However, the FEM calculations have also highlighted the significance of thermal residual stresses, which in many PMC and MMC

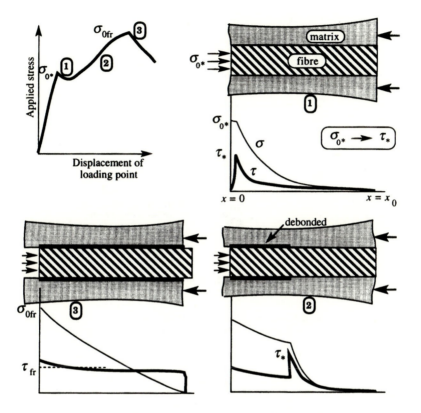

Fig. 7.6 Schematic stress distributions and load–displacement plot during the single-fibre push-out test. One difference from the pull-out test (see Fig. 7.4) is that the Poisson effect causes the fibre to expand (rather than contract), which raises (rather than offsets) the radial compressive stress across the interface due to differential thermal contraction.

systems will strongly affect the stress distributions. This effect, plus the errors introduced by using the simple analytical models, have probably been responsible for the rather poor agreement often observed between data from different tests. A variant of the basic push-out test has recently been developed (Kalton *et al.* 1994), involving externally applied tension in the plane of the specimen, which allows exploration of the effects of tensile stresses normal to the interface.

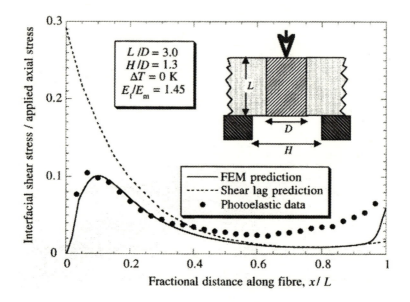

Fig. 7.7 Comparison between experimental (photoelastic) data and predictions from shear lag model and FEM modelling for the distribution of interfacial shear stress along the length of a fibre, starting from the top, during push-out of a resin 'fibre' in a matrix of a different resin (Kalton *et al.* 1994).

7.2.3 Other tests

A method which has been used for MMCs and some PMCs is the so-called 'full-fragmentation' technique. This procedure for deducing a shear strength involves embedding a single fibre in a matrix and straining the matrix in tension parallel to the fibre. The fibre fractures and fragments into a number of pieces. The aspect ratios exhibited by the resulting fibre segments are measured. Analysis is based on a constant τ, with the Weibull modulus of the fibre taken into account (see LePetitcorps *et al.* 1989). A related technique has been applied to thin ceramic coatings on a flat metallic substrate (Agrawal and Raj 1990).

Tests are sometimes carried out on complete laminae, rather than focusing on particular fibres. For example, Fig. 7.8 shows tests designed to measure intralaminar shear strength and transverse tensile strength. Such measurements may be useful (for example, they are required for the prediction of laminate failure – see §8.3), but it can be very difficult to relate the data obtained to actual fibre/matrix interfacial properties.

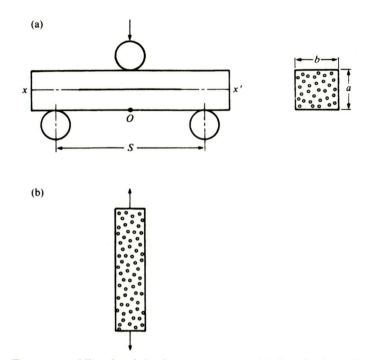

Fig. 7.8 Tests on unidirectional laminae to measure (a) intralaminar shear strength and (b) transverse tensile strength.

7.3 Control of bond strength

The interfacial bond strength can be influenced in a variety of ways. A brief summary is given below of the main types of effect involved.

7.3.1 Coupling agents and environmental effects

Many coatings have been developed to improve the durability and mechanical strength of the fibre–matrix bond and these are usually termed coupling agents. A good example is provided by those used on glass fibres, which often suffer from problems caused by pick-up of water. Some of the oxides in glass, such as SiO_2, Fe_2O_3 and Al_2O_3, form links to hydroxyl groups during contact with water and these in turn form hydrogen bonds to water molecules, so that glass picks up water very rapidly. In time, this can leach out other species in the glass, notably Na and Ca, to leave a weak, porous surface. The presence of the water also reduces the wettability of the fibres, reducing γ from 500–600 mJ m^{-2} to 10–20 mJ m^{-2}. In general, the interfacial shear strength falls as polymer composites are exposed to water, although with some thermoplastic matrices an increase has been observed (Gaur *et al.* 1994).

Coatings which function as coupling agents are designed to eliminate the leaching effect and raise the effective γ value at least to about 40–50 mJ m^{-2}. The primary function of the coupling agent is to provide a strong chemical link between the oxide groups on the fibre surface and the polymer molecules of the resin. A wide variety of commercial coupling agents have been developed (e.g. see Jones 1989), but the principles can be illustrated by the simple example shown in Fig. 7.9. This refers to the *silane coupling agents*, which have the general chemical formula R–Si–X$_3$. This is a multifunctional molecule which reacts at one end with the surface of the glass and at the other end with the polymer phase. The X units represent hydrolysable groups such as the ethoxy group (–OC$_2$H$_5$). The silane is hydrolysed to the corresponding silanol (see Fig. 7.9(a)) in the aqueous solution to which the fibres are exposed. These silanol molecules compete with water molecules to form hydrogen bonds with the hydroxyl groups bound to the fibre surface (Fig. 7.9(b)). When the fibres are dried, the free water is driven off and condensation reactions then occur, both at the silanol/fibre junction and between neighbouring silanol molecules (Fig. 7.9(c)). The result is a polysiloxane layer bonded to the glass surface, presenting an array of R groups to the environment.

This coating is water-resistant and can also form a strong bond to a polymer matrix. If the matrix is to be a thermosetting resin, then an R group is chosen which reacts with the resin during polymerisation, thus forming a permanent link. For a thermoplastic matrix, on the other hand, all the covalent links have been formed during manufacture of the polymer. However, choice of R with a fairly short chain which can interdiffuse with the chains of the matrix allows a strong bond to form. The efficiency of coupling agents in glass-fibre reinforced plastics (GFRP), and their effect on the behaviour of the composite, are evident in the fracture surfaces shown in Fig. 7.10. These are from composites made by injection moulding of polypropylene containing short glass fibres, with and without a coupling agent. It can be seen that the coupling agent promotes strong bonding, adhesion of a layer of plastic to the fibres and considerable plastic deformation of the matrix.

A further point to note about the function of such coupling agents is the possibility of bond formation being reversible. Plueddemann (1974)

(a)   R—Si X₃   +   H₂O   ⟶   R—Si (OH)₃   +   3HX

(b)

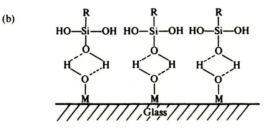

(c)

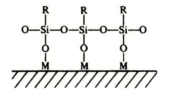

(d)

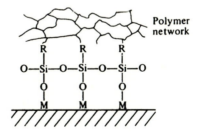

Fig. 7.9   Silane coupling agents: (a) hydrolysis of an organo-silane to the corre-sponding silanol, (b) hydrogen bonding between hydroxyl groups of the silanol and those attached to the glass surface, (c) polysiloxane bonded to the glass after condensation reactions during drying and (d) bonding between the functional group R and the polymer matrix.

proposed that movements at the interface could relax local stresses. This is shown in Fig. 7.11. In the presence of small quantities of water, sur-faces may be able to slide past each other without permanent bond fail-ure. Direct evidence for this reversible bonding has been obtained by Fourier Transform IR spectroscopy (Ishida and Koenig 1980).

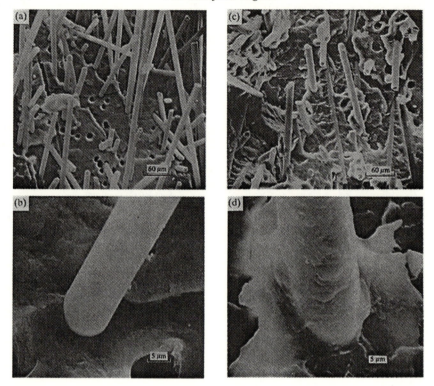

Fig. 7.10   SEM micrographs of fracture surfaces from injection moulded poly-propylene containing short glass fibres. A coupling agent was present on the fibres used to make specimens (c) and (d) but absent for (a) and (b). The higher bond strength induced by the coupling agent is clear from the adherence of polymer to the fibres and the shorter pull-out lengths. (Courtesy of Dow–Corning Corporation.)

For MMCs, it is less common for the promotion of good bonding to be necessary, because some local chemical reaction often occurs naturally during fabrication. There are, however, some systems in which wetting is very poor and coatings have been used to improve this (e.g. see Katzmann 1987). For CMCs, although various types of coating have been developed, these are rarely designed to improve wetting or adhesion. Fibres are normally added to ceramic matrices in order to improve the toughness and a relatively low debonding stress is usually preferred to promote frictional sliding during fibre pull-out (see §7.3.2).

(a)

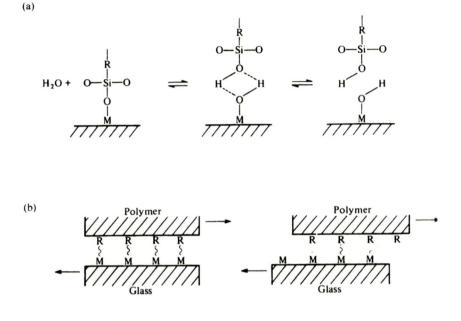

(b)

Fig. 7.11 (a) Mechanisms of reversible bond formation associated with hydrolysis as proposed by Plueddemann (1974). (b) Shear displacements without permanent damage to the interfacial bond.

7.3.2 Toughness-reducing coatings

It is sometimes preferable to ensure that the fibre/matrix interface has a low toughness, so that it debonds easily. This is usually the case when the main priority is to raise the toughness of the composite as a whole, by promoting crack-deflection at the interface, often leading to subsequent fibre pull-out by frictional sliding, which absorbs a substantial amount of energy (see §9.2). As an example of this, consider the data shown in Fig. 7.12. This shows the effect of carbon coatings, of increasing thickness, deposited on Nicalon™ fibres (§2.1.4) prior to incorporation in a SiC matrix. The interfacial shear strength is reduced progressively, but the macroscopic toughness increases. Graphite is popular for this purpose in ceramic systems and even thicker interface layers have been used in planar SiC laminates (see §9.2).

The promotion of interfacial debonding is often of over-riding importance for ceramic systems, but it can also be desirable in polymer composites, particularly with thermosetting matrices. Unfortunately,

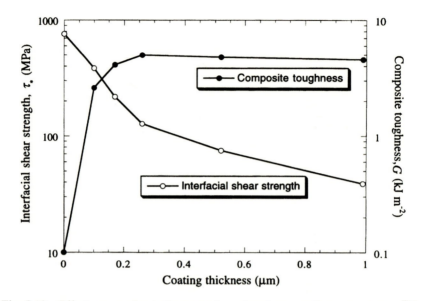

Fig. 7.12 Effect on mechanical properties of carbon coatings on Nicalon™ fibres in a SiC matrix. The plot is of measured interfacial shear strength (from push-down testing) and material toughness (from the area under the stress–strain curve during loading along the fibre axis) as a function of coating thickness (Lowden 1993).

interfaces which readily debond can lead to poor transverse and shear properties, and to poor resistance to environmental degradation such as water penetration. In MMCs, the contribution of fibre pull-out to the overall toughness is usually small and there is little incentive to encourage interfaces to debond readily.

7.3.3 *Interfacial chemical reaction and diffusion barrier coatings*

Interfacial reaction can be quite extensive in certain types of composite, particularly with metallic matrices. It may occur during composite fabrication and under in-service conditions. There are many fibre/matrix combinations for which chemical reaction is thermodynamically favoured. Extensive reaction is usually undesirable, since it tends to promote interfacial cracking and the reaction product itself is often a brittle ceramic or intermetallic compound (see Chapter 9). Techniques for avoiding excessive reaction depend either on relatively slow reaction kinetics or on the provision of some protective layer (diffusion barrier).

Interfacial chemical reactions are of particular concern for titanium composites. Titanium and its alloys tend to react with most reinforcements and there is interest in their use at elevated temperatures (550–700 °C). Furthermore, in the case of titanium, the surface oxide film which usually protects such reactive metals tends to dissolve in the matrix at temperatures above about 600 °C. Titanium reacts with virtually all reinforcement materials during fabrication, which requires temperatures of at least about 850 °C. In particular, quite substantial reaction occurs during fabrication with SiC monofilaments (Martineau *et al.* 1984), which is the most promising of the available long-fibre reinforcements for use in titanium. In general, fewer problems of interfacial reaction arise with aluminium, particularly during solid-state processing. Prolonged exposure of SiC to an aluminium melt does cause chemical reaction, but this is not a severe problem even for casting routes (see §11.2.2). Reaction problems can occur with magnesium alloys, although, since magnesium does not form a stable carbide, it is thermodynamically stable in contact with SiC. It does, however, tend to attack most oxides.

Coatings designed to protect the fibre against interfacial chemical attack have been extensively studied for MMCs, particularly titanium, and, to some extent, for CMCs. These coatings must be thermodynamically stable, with a low permeability to migrating reactants and a high resistance to mechanical damage. The first condition severely limits the choice of materials, while the second effectively determines a minimum layer thickness. Techniques such as Physical Vapour Deposition (PVD), Chemical Vapour Deposition (CVD) and sputter deposition tend to give coatings with a fine radial columnar grain structure, in which the grain boundaries provide good through-thickness diffusion paths.

In fact, most of the coatings developed hitherto for use in titanium MMCs are not thermodynamically stable in contact with Ti, but react relatively slowly and offer prolonged protection for the fibre itself. Carbon and duplex C/TiB_2 layers have been used. The micrograph in Fig. 7.13 shows the interfacial region of a Ti/SiC composite, with a C/TiB_2 coating on the fibre, after thermomechanical loading. In this case, some reaction occurred during fabrication to form the monoboride, TiB, and debonding took place between this layer and the titanium matrix.

7.3.4 *The interphase region*

The concept of an interfacial reaction can be generalised to encompass various modifications to the matrix microstructure in the vicinity of the

Fig. 7.13 SEM micrograph of a Ti/30% SiC monofilament composite after transverse loading with superimposed thermal cycling. The labelled regions are 1 – SiC fibre, 2 – C inner coating, 3 – TiB$_2$ outer coating, 4 – TiB reaction layer, 5 – Ti–6Al–4V matrix. (Courtesy of P. Feillard)

fibres. This idea has been extensively investigated for polymer composites. A summary of the effects involved has been presented by Hull (1994). Parts of the matrix which have been affected by the presence of the fibres are sometimes referred to as the *interphase* region. Since the microstructural modifications often induce changes in mechanical behaviour, the presence of the interphase can have a strong effect on the properties of the composite.

An example is provided by the micrograph shown in Fig. 7.14. This shows a carbon fibre/PEEK composite in which a degree of crystallisation has occurred in the matrix surrounding the fibres. Nucleation of spherulitic crystallites has occurred preferentially at the fibre surfaces. Such an increase in crystallinity tends to depress the fracture toughness and raise the tensile strength of the matrix. Furthermore, the volume contraction accompanying crystallisation will set up residual stresses which may affect the behaviour (see §7.1.5). Strong effects can also be produced in thermosetting matrices. It has been shown (Wright 1990) that the nature of the fibre surface can affect the curing kinetics and cross-link density of nearby matrix. Surface coatings of various types

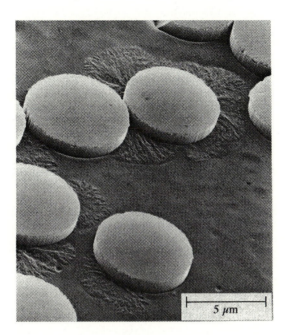

Fig. 7.14 SEM micrograph (Barlow *et al.* 1990) of an etched section from a carbon fibre/PEEK composite, showing partial crystallisation of the matrix around the fibres.

(§7.3.1) can modify the behaviour substantially. The proportion of the matrix affected in this way by proximity to the fibre surface can be substantial, particularly in composites with high fibre contents.

References and further reading

Agrawal, D. C and Raj, R. (1989) Measurement of the ultimate shear strength of a metal–ceramic interface, *Acta Metall.*, **37** 1265–70

Barlow, C. Y., Peacock, J. A. and Windle, A. H. (1990) Relationships between microstructures and fracture energies in carbon-fibre/PEEK composites, *Composites*, **21** 383–8

Cappleman, G. R., Watts, J. F. and Clyne, T. W. (1985) The interfacial region in squeeze infiltrated composites containing δ-alumina fibre in an aluminium matrix, *J. Mat. Sci.*, **20** 2159–68

Chua, P. and Piggott, M. R. (1985) The glass fibre–polymer interface: I Theoretical considerations for single fibre pullout tests, *Comp. Sci. Tech.*, **22** 33–42

Clegg, W. J., Horsefall, I., Mason, J. F. and Edwards, L. F. (1988) The tensile deformation and fracture of Al–Saffil^{TM} metal matrix composites, *Acta Metall.*, **36** 2151–9.

Evans, A. G. and Dalgleish, B. J. (1993) The fracture resistance of metal–ceramic interfaces, *Mat. Sci. & Eng.*, **162A** 1–13

Evans, A. G., Ruhle, M., Dalgleish, B. J. and Charalambides, P. G. (1990) The fracture energy of bimaterial interfaces, *Mat. Sci. & Eng.*, **126A** 53–64

Favre, J. P. (1989) Review of test methods and testing for assessment of fibre–matrix adhesion, in *Interfacial Phenomena in Composite Materials 1989*. F. R. Jones (ed.) Butterworths: London pp. 282–93

Gaur, U., Chou, C. T. and Miller, B. (1994) Effect of hydrothermal ageing on bond strength, *Composites*, **25** 609–12

Grande, D. H., Mandell, J. F. and Hong, K. C. C. (1988) Fibre–matrix bond strength studies of glass, ceramic and metal matrix composites, *J. Mat. Sci.*, **23**, 311–28

Hsueh, C. H. (1990) Evaluation of interfacial shear strength, residual clamping stress and coefficient of friction for fibre-reinforced ceramic composites, *Acta Metall. Mater.*, **38** 403–9

Hull, D. (1994) Matrix-dominated properties of polymer matrix composite materials, *Mater. Sci. & Eng.*, **A184** 173–83

Ishida, H. and Koenig, J. L. (1980) Hydrolytic stability of silane coupling agents on E-glass studied by Fourier Transform infra-red spectroscopy, *Proc. 35th SPI/RP Ann. Tech. Conf.*, paper 23-A, Soc. Plas. Ind.

Jones, F. R. (1989) Interfacial aspects of glass fibre reinforced plastics, in *Interfacial Phenomena in Composite Materials 1989*. F. R. Jones (ed.) Butterworths: London pp. 25–32

Kallas, M. N., Koss, D. A., Hahn, H. T. and Hellman, J. R. (1992) Interfacial stress state present in a 'thin slice' fiber push-out test, *J. Amer. Ceram. Soc.*, **74** 1585–96

Kalton, A. F., Ward-Close, C. M. and Clyne, T. W. (1994) Development of the tensioned push-out test for study of fibre/matrix interfaces, *Composites*, **25** 637–44

Katzman, H. A. (1987) Fibre coatings for the fabrication of graphite reinforced magnesium composites, *J. Mat. Sci.* **22** 144–8

Kendall, K. (1985) Fracture mechanics of interface failure, *Mat. Res. Bull.*, **40** 167–76

Kerans, R. J. and Parthasarathy, T. A. (1991) Theoretical analysis of the fiber pullout and pushout tests, *J. Mat. Sci.*, **27** 3821–6

Lawrence, P. (1972) Some theoretical considerations of fibre pullout from an elastic matrix, *J. Mat. Sci.*, **7** 1–6

LePetitcorps, Y., Pailler, R. and Naslain, R. (1989) The fibre/matrix interfacial shear strength in titanium alloy matrix composites reinforced by SiC or B CVD filaments, *Comp. Sci. & Tech.* **35** 207–14

Lowden, R. A. (1993) Fiber coatings and the mechanical properties of a continuous fiber reinforced SiC matrix composite, in *Designing Ceramic Interfaces II*. S. D. Peteves (ed.) Comm. of Europ. Communities: Luxembourg pp. 157–72

Marshall, D. B. (1985) An indentation method for measuring fibre–matrix frictional stresses in ceramic composites, *J. Amer. Ceram. Soc.*, **67** 259–60

Marshall, D. B. and Oliver, W. C. (1987) Measurement of interfacial mechanical properties in fiber-reinforced ceramic composites, *J. Amer. Ceram. Soc.*, **70** 542–8

Martineau, P., Lahaye, M., Pailler, R., Naslain, R., Couzi, M. and Cruege, F. (1984) SiC filament/titanium matrix composites regarded as model composites. Part II: Fibre/matrix reactions at high temperatures, *J. Mat. Sci.*, **19** 2731–48

Mikata, Y. and Taya, M. (1985) Stress field in a coated continuous fibre composite subjected to thermomechanical loadings, *J. Comp. Mat.*, **19** 554–79

Plueddemann, E. P. (1974) Mechanisms of adhesion through silane coupling agents, in *Composite Materials, Vol. 6*. E. P. Plueddemann (ed.) Academic: New York pp. 174–216

Scola, D. A. (1974) High-modulus fibres and the fibre–resin interface in resin composites, in *Composite Materials, Vol. 6*. E. P. Plueddemann (ed.) Academic: New York pp. 217–84

Takehashi, H. and Chou, T. W. (1988) Transverse elastic moduli of unidirectional fibre composites with interfacial debonding, *Metall. Trans.*, **19A** 129–35

Watson, M. C. and Clyne, T. W. (1992) The use of single fibre push-out testing to explore interfacial mechanics in SiC monofilament reinforced Ti, *Acta Metall. Mater.*, **40** 135–45

Wright, W. W. (1990) The carbon fibre/epoxy resin interface – a review, *Compos. Polym.*, **3** 231–57

8

Strength of composites

The elastic behaviour of long- and short-fibre composites was described in Chapters 4 to 6. The stresses in the individual plies of a laminate under an external load and the stress distributions along short fibres were examined. This information is used to explore the ways in which a material suffers microstructural damage, leading to the ultimate failure of a component. There are two important aspects to this behaviour. Firstly, there is the deflection, degree of damage and ultimate failure of a component as a function of applied load. Secondly, there are the processes which cause absorption of energy within a composite material as it is strained. The latter determine the toughness of the material and are treated in Chapter 9. In the present chapter, attention is concentrated on predicting the applied stress at which damage and failure occur. The treatment is oriented towards long-fibre materials and laminates, and, in particular, towards polymer-based composites. Most of the principles apply equally to discontinuous reinforcement and other types of matrix. Some specific points concerning failure of such systems are dealt with in Chapter 9.

8.1 Failure modes of long-fibre composites

The application of an arbitrary stress state to a unidirectional lamina can lead to failure by one or more basic failure processes. The three most important types of failure are illustrated in Fig. 8.1. Large tensile stresses parallel to the fibres, σ_1, lead to fibre and matrix fracture, with the fracture path normal to the fibre direction. The strength is much lower in the transverse tension and shear modes and the composite fractures on surfaces parallel to the fibre direction when appropriate σ_2 or τ_{12} stresses are applied. In these cases, fracture may occur entirely within the matrix,

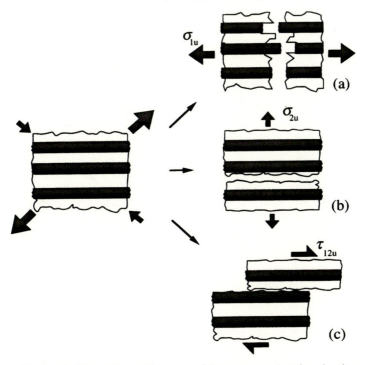

Fig. 8.1 Schematic illustration of how an arbitrary stress state in a lamina gives rise to failure as a result of exceeding critical values of (a) axial tensile stress σ_{1u}, (b) transverse tensile stress σ_{2u} and (c) shear stress τ_{12u}.

at the fibre/matrix interface or primarily within the fibre. To predict the strength of a lamina or laminate, values of the failure (ultimate) stresses σ_{1u}, σ_{2u} and τ_{12u} have to be determined.

8.1.1 Axial tensile failure

Understanding of failure under an applied tensile stress parallel to the fibres is relatively simple, provided that both constituents behave elastically and fail in a brittle manner. They then experience the same axial strain and hence sustain stresses in the same ratio as their Young's moduli. Two cases can be identified, depending on whether matrix or fibre has the lower strain to failure, as illustrated in Fig. 8.2(a) and (b).

In case (a), the matrix has the lower failure strain ($\epsilon_{mu} < \epsilon_{fu}$). For strains up to ϵ_{mu}, the composite stress is given by the simple rule of mixtures

$$\sigma_1 = f\sigma_f + (1 - f)\sigma_m \qquad [(4.2)]$$

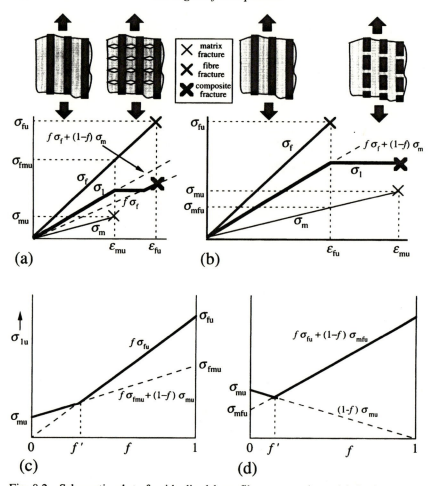

Fig. 8.2 Schematic plots for idealised long-fibre composites with both components behaving in a brittle manner. (a) and (c) refer to a system in which the fibre has a higher strain to failure than the matrix and show respectively stress–strain relationships (of fibre, matrix and composite) and dependence of composite failure stress on volume fraction of fibre. (b) and (d) show the same plots for the case where the matrix has the higher strain to failure.

Above this strain, however, the matrix starts to undergo microcracking and this corresponds with the appearance of a 'knee' in the stress–strain curve, as shown in Fig. 8.2(a). The composite subsequently extends with little further increase in the applied stress. As matrix cracking continues, the load is transferred progressively to the fibres. If the strain does not reach ϵ_{fu} during this stage, further extension causes the composite stress to rise and the load is now carried entirely by the fibres. Final fracture occurs when the strain reaches ϵ_{fu}, so that the composite failure stress σ_{lu} is given by $f\sigma_{fu}$.

Alternatively, if the fibres break before matrix cracking has become sufficiently extensive to transfer all the load to them, then the strength of the composite is given by

$$\sigma_{lu} = f\sigma_{fmu} + (1 - f)\sigma_{mu} \tag{8.1}$$

where σ_{fmu} is the fibre stress at the onset of matrix cracking ($\epsilon_1 = \epsilon_{mu}$). The composite failure stress depends therefore on the fibre volume fraction in the manner shown in Fig. 8.2(c). The fibre volume fraction above which the fibres can sustain a fully transferred load is obtained by setting the expression in Eqn (8.1) equal to $f\sigma_{fu}$, leading to

$$f' = \frac{\sigma_{mu}}{\sigma_{fu} - \sigma_{fmu} + \sigma_{mu}} \tag{8.2}$$

In case (b), shown in Fig. 8.2(b) and (d), $\epsilon_{mu} > \epsilon_{fu}$. The fibres fail first, at a composite strain of ϵ_{fu}. Further straining causes the *fibres* to break up into progressively shorter lengths and the load to be transferred to the *matrix*. This continues until all the fibres have aspect ratios below the critical value (see §6.1.5). It is often assumed in simple treatments that only the matrix is bearing any load by the time that break-up of fibres is complete. Subsequent failure then occurs at an applied stress of $(1 - f)\sigma_{mu}$. If matrix fracture takes place while the fibres are still bearing some load, as shown in Fig. 8.2(b), then the composite failure stress is

$$\sigma_{lu} = f\sigma_{fu} + (1 - f)\sigma_{mfu} \tag{8.3}$$

where σ_{mfu} is the matrix stress at the onset of fibre cracking. In principle, this implies that the presence of a small volume fraction of fibres reduces the composite failure stress below that of the unreinforced matrix, as shown in Fig. 8.2(d). This occurs up to a limiting value f' given by setting the right-hand side of Eqn (8.3) equal to $(1 - f)\sigma_{mu}$

$$f' = \frac{\sigma_{mu} - \sigma_{mfu}}{\sigma_{fu} - \sigma_{mfu} + \sigma_{mu}} \tag{8.4}$$

An example of the application of this analysis to real systems is given in Table 8.1, for two polymer matrix composites and one ceramic matrix composite. (It is unrealistic to apply this calculation to ductile metallic and thermoplastic matrices, which normally depart from linear elastic behaviour well before failure occurs.) For the two cases where matrix failure occurs first, the volume fraction, f', above which the fibres can

Table 8.1 *Tensile failure data, based on linear elastic behaviour*

Composite system	Young's modulus (GPa)	Failure strain (%)	Failure stress (MPa)	σ_{mfu} (MPa)	σ_{fmu} (MPa)	f' (vol. %)	σ_{lu} for $f = 50\%$ (MPa)
Type 1 (HM) C fibre	350	0.7	2500	35	—	2.5	1267
Epoxy matrix	4	2.0	100				
Glass fibre	71	2.8	2000	—	1430	10.9	1000
Polyester matrix	3	2.0	70				
Nicalon™ fibre	130	0.8	1000	—	175	10.8	500
Glass matrix	70	0.14	100				

Data for the matrices and fibres are typical experimental values. Three composite systems are being considered, formed by taking the listed fibres and matrices in pairs. The parameters σ_{mfu}, σ_{fmu}, f' and σ_{lu} refer to the failure behaviour of these composites under axial tensile loading; they are defined by the plots in Fig. 8.2. Their values have been calculated from the data for the fibres and matrices.

carry all the load is about 11%. With the carbon/epoxy system, the fibres make a contribution to the failure stress at fibre contents above about 2.5%. Since most long-fibre composites have fibre volume fractions between about 30% and 70%, it is evident that the fibre strength is dominant in determining the axial strength of the composite and the product $f\sigma_{fu}$ provides a reliable estimate of σ_{lu} (at least for polymer matrix composites).

Although the above conclusion about estimation of σ_{lu} is often valid, the overall treatment represents a gross simplification. In reality, micro-cracking of the matrix does not result in the matrix becoming completely unloaded and fibres still carry some stress even after they have broken into short lengths. After the onset of damage at the 'knee', there is a change in the slope of the stress–strain curve, but it does not in fact reduce to zero (cf. Fig. 8.2(a)). These effects arise because load is trans-ferred across the interface even after the fibre or matrix fractures.

Further assumptions in the above treatment are that the fibre strength is a constant and that the fibres fail in isolation from each other. In fact, most types of fibre exhibit a range of strengths (see §2.2.4); the variability in strength is greater when the fibre displays a low Weibull modulus. Thus, when a load is applied parallel to the fibres, the first failure is at the weakest point. If the stress redistribution associated with this failure is not sufficient to cause adjacent fibres to break, the applied stress increases and further fractures occur randomly throughout the material, as illustrated in Fig. 8.3. Several models have been proposed to treat this process, broadly falling into two groups. In the **cumulative weakening models**, first proposed by Rosen (1965), a random sequence of fibre fractures occurs with increasing load until the residual strength of the composite across a section somewhere along the length is reached. This provides an upper bound on the strength, since in practice there is a tendency for damage to become concentrated in certain regions. In **fibre break propagation models**, suggested by Zweben and Rosen (1970), the initial failure sequence again involves fracture of individual fibres at weak points. As each fibre breaks, redistribution of stress occurs, leading to additional stresses on neighbouring fibres associated with a local stress magnification effect. Thus, there is an increased probability that fracture will occur in closely adjacent fibres. This is illustrated by the stress distribution plots in Fig. 8.3, which show peaks where neighbour-ing fibres have broken.

The models have limited predictive power, because the stress field around a fractured fibre is sensitive to details of interfacial structure (see

F

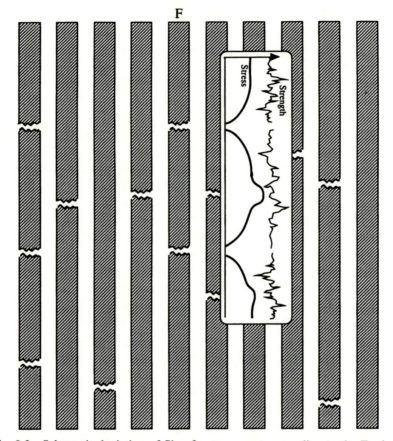

Fig. 8.3   Schematic depiction of fibre fracture events according to the Zweben–Rosen model of tensile failure in aligned long-fibre composites. The distribution of axial stress, and of local fibre strength (due to the presence of flaws) along the length is shown for the fibre marked F.

Chapter 7) and to matrix yielding and fracture behaviour, which can be influenced by a wide variety of factors. However, some expected trends can be identified. Clearly, the relative importance of cumulative weakening and fibre break propagation depends on the Weibull modulus and the interfacial properties. For example, fibre break propagation becomes more important when the Weibull modulus is high, while randomly distributed fracture events predominate when there is a wide variation in local fibre strength, represented by a low Weibull modulus. It may also be noted that the failure strength of the composite is expected to show much less variability, and sensitivity to specimen size, than do the constituent individual fibres (see, for example, Batdorf and Gaffarian 1984).

Prediction of stress fields around cracks by the use of numerical modelling has proved useful in exploring what is likely to occur after a fibre has fractured. Pioneering work of this type was carried out by Cook and Gordon (1964). They presented stress magnification contours around a crack tip, which are shown in Fig. 8.4. This demonstrates that, while the σ_1 stress parallel to the applied load is substantially magnified near the crack tip (Fig. 8.4(a)), there is also a significant transverse σ_2 stress, in the direction of crack propagation (Fig. 8.4(b)). This transverse stress, which peaks at a point slightly in front of the crack tip, may cause debonding of a fibre/matrix interface in front of the crack, as shown in Fig. 8.4(e). This can open up the interface and thus blunt the crack tip, tending to prevent

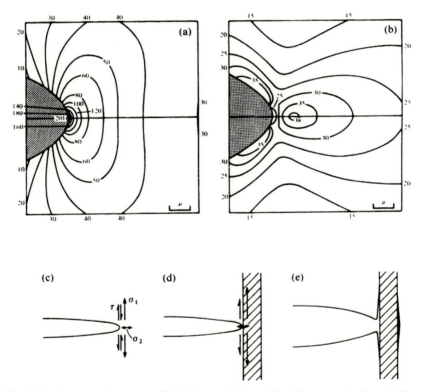

Fig. 8.4 Stresses close to an elliptical crack tip predicted by numerical modelling (Cook and Gordon 1964). (a) Contours of σ_1 stress, as a ratio to the applied stress. (b) Contours of σ_2 stress, in the transverse direction, as a ratio to the applied stress. (c) Geometry of σ_1, σ_2 and τ stresses. (d) Crack tip approaching the fibre/matrix interface. (e) Interfacial debonding as a result of the σ_2 stress, blunting the crack.

the crack from simply slicing through the fibres it encounters in the crack plane. This ***crack-blunting mechanism*** is important in raising the toughness of the composite. Criteria for such crack deflection are examined in §9.1.2.

Cook and Gordon's modelling was done for an isotropic, elastic continuum. More recent work has explored the significance of fibre/matrix stiffness mismatch, plastic flow in the matrix, interfacial sliding, etc. For example, the work of He *et al.* (1993) is based on the geometrical model shown in Fig. 8.5. Predictions from this model are shown in Fig. 8.6. In

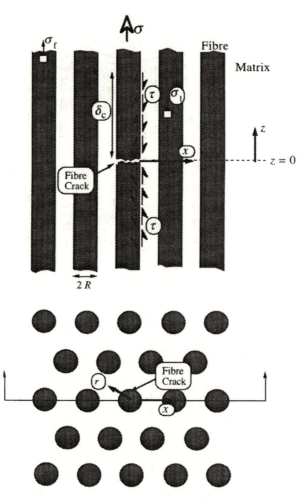

Fig. 8.5 Schematic depiction of the geometrical arrangement devised by He *et al.* (1993) for numerical modelling of the stress field around a cracked fibre in an hexagonal array, under axial load.

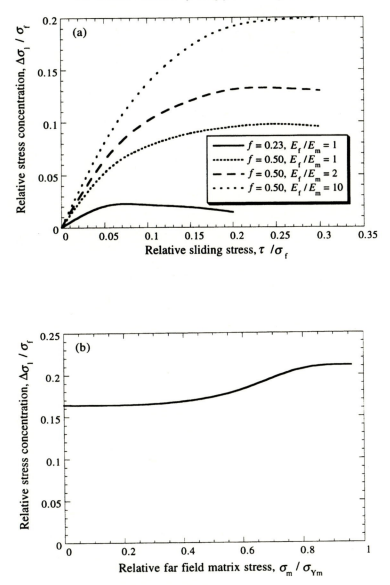

Fig. 8.6 Predictions from the model of He *et al.* (1993). The extra increment $\Delta\sigma_1$ of fibre stress in the direction of applied load, due to the presence of the crack in the neighbouring fibre, is plotted as a proportion of the stress carried by the fibres remote from the crack σ_f. (a) Effect of the shear stress τ for interfacial sliding adjacent to the crack. (b) Effect of the ratio between the matrix stress remote from the crack and the matrix yield stress, for the case of $E_f/E_m = 3$, $f = 50\%$ and $\tau/\sigma_f = 0.2$.

Fig. 8.7   SEM micrograph of the fracture surface of an epoxy/60% carbon fibre
lamina tested under axial tension. The fracture surface is relatively smooth and
consists of a network of blocky outcrops of fibres and resin at different levels.

Fig. 8.6(a), the peak stress in a fibre neighbouring the cracked one is
plotted as a function of the interfacial shear stress, for two fibre
volume fractions and three fibre/matrix stiffness ratios. Several effects
are apparent here. Firstly, the stress concentration is greater for a
higher ratio of fibre stiffness to matrix stiffness. Secondly, stress concentration becomes negligible as the interfacial shear strength falls to
zero; this is expected, since the stress concentrating effect of the crack
is lost if stresses cannot be transmitted to the adjacent matrix. Finally,
the stress concentration is reduced if the neighbouring fibre is further
away (lower fibre content). A further effect is apparent in Fig. 8.6(b).
This shows the effect of plastic flow occurring in the matrix. At higher
values of the ratio between the matrix stress and its yield stress (i.e.
more extensive yielding), the stress concentration increases somewhat.
This is expected, since yielding has a similar effect to decreasing the
matrix stiffness.

The practical effect of changing the interfacial shear strength is illustrated in Figs. 8.7–8.10, which show various fracture surfaces from long-

Fig. 8.8 SEM micrograph of the fracture surface of a polyester/60% glass fibre lamina tested under axial tension. Extensive fibre pull-out has occurred.

fibre composites subjected to axial tension. For the epoxy/carbon fibre composite shown in Fig. 8.7, the interfacial bond strength was high, causing strong stress concentration and a tendency for cracks to pass through large bundles of fibres without much deviation. For the polyester/glass and epoxy/Kevlar^{TM} composites shown in Figs. 8.8 and 8.9, on the other hand, the bonding was weaker and extensive fibre pull-out has occurred. Also evident in Fig. 8.9 is the tendency for Kevlar^{TM} fibres to undergo necking and fibrillation (see §2.1.3). Finally, Fig. 8.10 shows that the presence of an adverse environment (in this case hydrochloric acid), which penetrates along the advancing crack, can dramatically reduce the fibre strength (relative to the interfacial strength) and hence

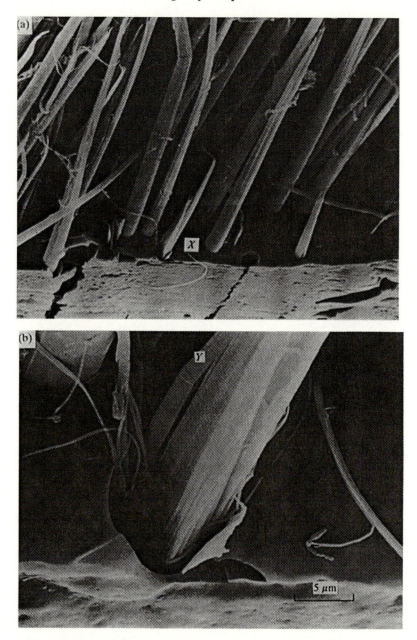

Fig. 8.9 SEM micrograph of the fracture surface of an epoxy/40% Kevlar™ 49 fibre lamina tested under axial tension. (b) is a higher magnification view of the fibre marked X in (a) and shows fibrillation and kink band formation in the fibre at Y.

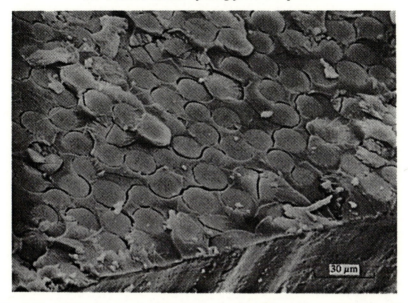

Fig. 8.10   SEM micrograph of the fracture surface of a polyester/60% glass fibre lamina tested under axial tension in the presence of dilute hydrochloric acid.

cause a highly planar fracture with a much reduced failure stress. The important question of the energy absorption associated with different types of failure is examined in Chapter 9.

For metal-based composite laminae, the situation is rather different because the matrix usually has the capacity to deform plastically to large strains. A diagram analogous to Fig. 8.2(b) can be constructed, taking account of the matrix work-hardening behaviour. Fig. 8.11 shows (a) a schematic stress–strain curve and (b) predicted and experimental dependence of failure stress on fibre volume fraction for composites made up of tungsten wires in copper. The experimental data are in good agreement with simple theory for this system.

8.1.2 Transverse tensile failure

It is not possible to make a simple estimate of σ_{2u}, comparable with the estimate of σ_{1u} described in the previous section. The transverse strength is influenced by many factors, such as the nature of the interfacial bonding, the fibre distribution, the presence of voids, etc. In general, the

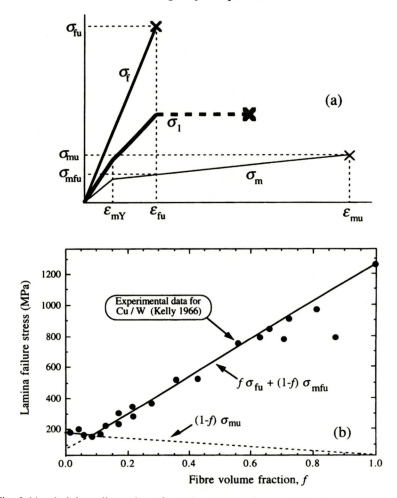

Fig. 8.11   Axial tensile testing of metal matrix laminae. (a) Idealised stress–strain plots for fibre, matrix and composite, and (b) a comparison between theory and experiment (Kelly and Macmillan 1986) for the dependence of failure stress on fibre volume fraction.

strength is less than that of the unreinforced matrix, often significantly so, and the strain to failure can be even more dramatically reduced. A consequence of this is that the transverse plies in a crossply laminate usually start to crack before the parallel plies, even though they are less stiff and so carry less load – see §8.3.

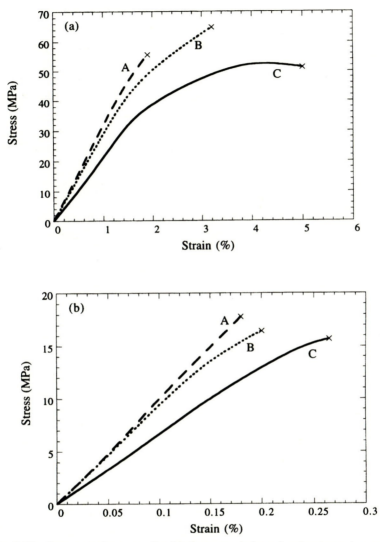

Fig. 8.12 Stress–strain curves for (a) three unreinforced polyester resins and (b) laminae based on these resins, with 48% glass fibre, tested in transverse tension (Legg 1980).

The influence of fibres on the transverse strength is illustrated by the experimental data shown in Fig. 8.12, which compares transverse tensile stress–strain plots for laminae based on three polyester resins with the behaviour of these materials in the unreinforced form. Both the strength and the strain to failure have been markedly reduced by the presence of the fibres. This is largely due to the inherent tendency for high local stresses and strains to develop in the matrix (see Fig. 4.3(b)). The fibres make little contribution to the strength. If the interfacial bonding is weak, then cracks tend to form at the interface and link up through highly stressed sections of matrix. A process of this type is illustrated in the micrographs of Fig. 8.13. If, on the other hand, there is strong resistance to interfacial decohesion, cracks will tend to form in the matrix close to the interface – where there is a concentration of stress and a high degree of triaxial constraint, so that matrix plasticity is inhibited. Alternatively, some fibres (such as carbon fibres) with a layered structure have little transverse strength and may fail internally.

With a metallic matrix, broadly similar characteristics are exhibited. Plots are shown in Fig. 8.14 which illustrate the effect of interfacial bond strength in a titanium composite. These are **Poisson plots**, in which the strain transverse to the loading direction is shown as a function of that parallel to the applied load. The gradient gives the Poisson ratio. The plots are for axial and transverse loading, before and after a heat treatment which raised the interfacial bond strength. Under axial loading, the Poisson's ratio rises slightly as inelastic behaviour starts; this is due to plastic flow in the matrix (with an associated Poisson's ratio of 0.5). Under transverse loading, however, there is a tendency for the Poisson's ratio to fall; this is a result of the interface opening up, which allows extension in the loading direction with little lateral contraction. After the heat treatment, this effect is very limited and failure occurs at a much reduced strain.

An estimate of the effect of the presence of the fibres on the transverse strength can be obtained by treating the fibres in the composite as a set of cylindrical holes and considering the reduction in load-bearing cross-section thus introduced. (This is not accurate, since the presence of even completely debonded fibres would lead to a different stress distribution in the matrix than would be the case for holes, but the approach is useful as a guide.) For a simple square array of holes, consideration of the maximum reduction in matrix cross-sectional area leads to the following expression for the transverse strength of a lamina having a volume fraction f of fibres

$$\sigma_{2u} = \sigma_{mu} \left[1 - 2 \left(\frac{f}{\pi} \right)^{1/2} \right] \tag{8.5}$$

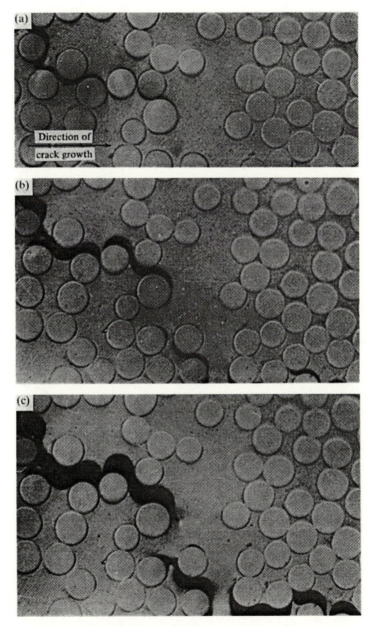

Fig. 8.13 SEM micrographs illustrating the propagation of a transverse crack in a polyester/glass lamina (Jones 1981).

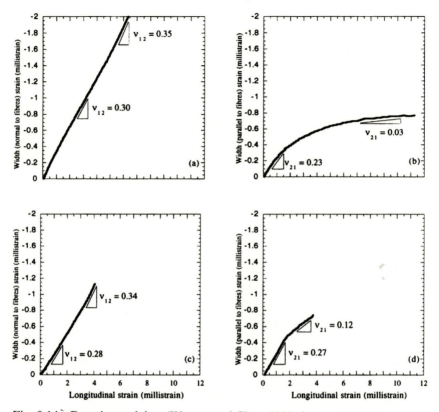

Fig. 8.14 Experimental data (Watson and Clyne 1993) from axial and transverse tensile testing of Ti–6Al–4V/SiC monofilament laminae, plotted in the form of lateral contraction as a function of extension in the loading direction. Plots are for (a) axial and (b) transverse loading of as-fabricated composite, and for (c) axial and (d) transverse loading after a heat treatment which removed the graphitic coating on the fibres.

A comparison is shown in Fig. 8.15 between predictions from this equation and experimental data for Al(6061)/Borsic composites. In this case, the model gives good agreement with experiment.

There is interest in improving the transverse properties of laminae, particularly the failure strain. Among the possibilities that have been considered is the provision of a very compliant (e.g. rubber) layer on the fibre surface, so as to reduce the strain localisation and constraint imposed on the matrix. Unfortunately, a layer sufficiently thick to increase significantly the transverse failure strain tends to have an adverse effect on other properties, such as the shear stiffness.

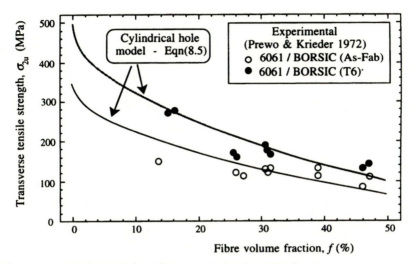

Fig. 8.15   Experimental data (Prewo and Kreider 1972) for the transverse tensile strength of Al/BORSIC laminae as a function of fibre volume fraction, compared with predictions from Eqn (8.5).

8.1.3 Shear failure

As with tensile fracture, shear failure tends to occur on planes determined by the fibre direction. The six possible combinations of plane and shearing direction, and their indices, are depicted in Fig. 8.16. There are three sets of equivalent pairs. Normally, there is considerable resistance to the fracture of fibres, so that the pair of modes denoted τ_{21} and τ_{31} are unlikely to occur. Of the other two pairs, involving sliding of fibres over one another either axially (τ_{12}) or laterally (τ_{32}), it is not obvious whether one is inherently more likely than the other. However, when considering the stressing of a thin lamina in the 1–2 plane, stresses of the τ_{32} type do not arise and only the magnitude of τ_{12u} is important. Broadly speaking, this is affected by the same factors as the transverse tensile strength, because shear stresses and strains become concentrated in the matrix between fibres in a manner similar to that outlined above for tensile stresses and strains. However, the details of this dependence are different. There is more scope for local matrix deformation to take place (without cracking) under this type of stress and local stress concentrations are relaxed more readily.

No simple analytical expression is available to predict the effect of fibre content on τ_{12u}. Adams and Dorner (1967) have used finite difference

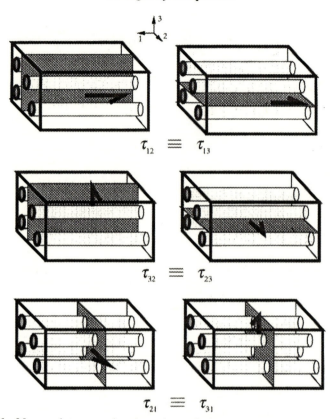

Fig. 8.16   Nomenclature and orientation of shear stresses acting within an
aligned fibre composite.

methods to deduce how the shear stress concentration factor should vary
with fibre volume fraction. The results are shown in Fig. 8.17. Unless the
fibre volume fraction is very high (when constraint on matrix deformation becomes severe), this factor is quite close to unity and τ_{12u} is expected
to have a value close to τ_u for the matrix. This is broadly confirmed by
the experimental data summarised in Table 8.2. It may be noted that
there are both practical and theoretical difficulties to be overcome in
obtaining such data. These are outlined in §8.2.3.

8.1.4 Failure in compression

Failure in *compression* is dependent on the way that the loading is applied
and, in particular, on the degree of lateral constraint. Under axial

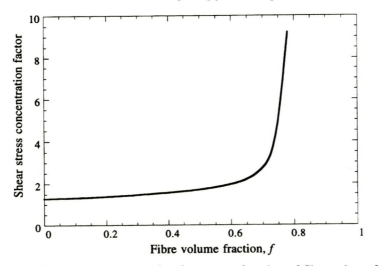

Fig. 8.17   Shear stress concentration factor as a function of fibre volume fraction, predicted by finite difference modelling for a square array of fibres (Adams and Dorner 1967).

Table 8.2 *Typical experimental failure data for laminae based on thermoset matrices*

Composite	Axial strength σ_{lu} (MPa)	Transverse strength σ_{2u} (MPa)	Shear strength τ_{12u} (MPa)	Axial failure strain ϵ_{lu} (%)	Transverse failure strain ϵ_{2u} (%)
Polyester/50% glass	700	20	50	2.0	0.3
Epoxy/50% carbon (HM)	1000	35	70	0.5	0.3
Epoxy/50% Kevlar^{TM}	1200	20	50	2.0	0.4

compression, there is a tendency for the fibres to **buckle**. Except at low fibre volume fractions, neighbouring fibres are constrained to buckle in phase, as shown in Fig. 8.18. Buckling results in compressive and tensile stresses across different parts of the fibre section, leading either to fracture or (if the fibre is one, such as Kevlar^{TM}, which can deform significantly) to local distortion. If buckling becomes extensive, then it will cause general collapse, i.e. failure, of the specimen.

A more common type of failure occurs from the onset of a local buckling instability. A **kink band** of misoriented fibres may form, as illustrated in Fig. 8.19. Figure 8.20 shows a fractured carbon fibre reinforced

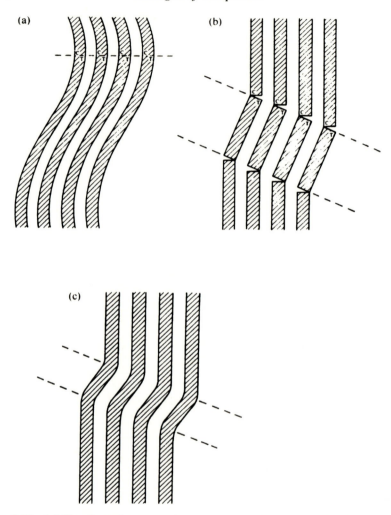

Fig. 8.18 (a) Tensile and compressive stresses in a fibre due to in-phase buckling, leading to a kink zone. (b) Two planes of fracture formed with brittle carbon fibres. (c) Unfractured kink zone formed with Kevlar™ 49 fibres.

composite which has failed in this way in the compression zone of a specimen loaded in four point bending. Jelf and Fleck (1992) have shown that most failures under axial compression are of this type, which they refer to as **plastic microbuckling**. Plastic deformation of the matrix is necessary for the mechanism to initiate. The main factors which influence the onset of this type of instability, apart from fibre content, are matrix shear yield stress, τ_{Ym}, and (average) fibre misalignment angle,

Fig. 8.19 Optical micrograph of a polished section of a carbon fibre/epoxy resin pultruded specimen, showing a kink band formed in the compression zone of a four-point bend test specimen (Parry and Wronski 1981).

$\Delta\phi$. Argon (1972) proposed that the compressive failure stress could be expressed as

$$\sigma_{c^*} = \frac{\tau_{Ym}}{\Delta\phi}$$ (8.6)

where $\Delta\phi$ is in radians. It is assumed that the volume fraction of fibres is high enough for this type of failure to be likely. In this regime, the failure stress is predicted to be independent of fibre content. It is also assumed that the interfacial bond strength is high enough to ensure that debonding does not initiate failure.

The data shown in Fig. 8.21, taken from a number of different studies, confirm that there is a strong correlation between matrix shear yield stress and composite failure stress in compression. The gradient of the best-fit line in this plot corresponds, via Eqn (8.6), to a value for $\Delta\phi$ of about 3°. This is physically reasonable, in that measured misalignments in nominally unidirectional composites often range up to about this figure. Errors in specimen gripping geometry might typically involve misalignment up to $\sim 1°$. Furthermore, there is some evidence

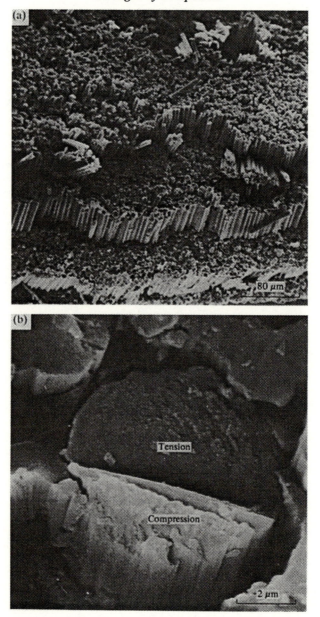

Fig. 8.20 SEM micrograph of the fracture surface of a carbon fibre/epoxy resin lamina after a fibre buckling mode failure due to a longitudinal compressive stress. (a) Low magnification view showing smooth fracture surface. (b) High-magnification view showing tension and compressive fracture in a single fibre. (Ewins and Potter 1980).

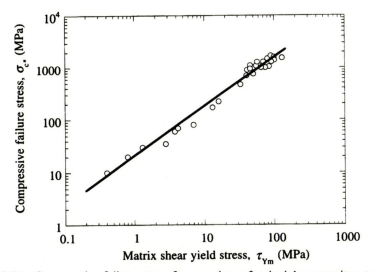

Fig. 8.21 Compressive failure stress for a variety of uniaxial composites, taken from several different studies, plotted against the shear yield stress of the matrix (Jelf and Fleck 1992).

that composites showing particularly good alignment have excellent compressive strengths. It is, however, difficult to be precise about the reliability of Eqn (8.6), since it is unclear how large a group of fibres needs to be misaligned by $\Delta\phi$ in order for kink-band formation to initiate. This is an area requiring further research.

A composite can also fail under axial compression by macroscopic shear on certain planes. This must involve fracture or gross deformation of fibres, since no shear stresses are developed on any planes parallel to the fibre axis. This usually occurs at applied loads similar to those necessary to cause axial tensile failure, although modelling of the mechanisms involved is difficult. These compressive failure loads can be significantly reduced if *fibre buckling* occurs. This is generally favoured by measures which reduce the stiffness of the matrix, such as heating or, with some polymer matrices, prolonged exposure to water. Reductions in the interfacial bond strength (§7.3) also tend to facilitate this. Finally, there is a significant dependence on the fibre diameter. Large-diameter fibres are much more resistant to buckling than fine fibres. This point was made in §2.2.2. The sensitivity to diameter can be inferred from the Euler buckling formula given as Eqn (2.1) and the fibre flexibility data in Table 2.4. An example of exploitation of the excellent resistance of large-diameter fibres to buckling is given in §12.4, relating to the use of boron monofilaments in golf clubs.

In general, the stress–strain plot of a composite loaded under axial compression is rather similar to the corresponding tensile plot. Commonly, the initial gradient is slightly reduced and the failure stress somewhat lower under compression. These changes reflect the straining and damage development which can occur under compression when fibres become misaligned. These compression tests become meaningless if macroscopic (Euler) buckling of the specimen is allowed to occur. It is therefore necessary to ensure that the specimen has a low aspect ratio (length/diameter) and/or that suitable anti-buckling guides are used. Application of the technique to thin laminates is described by Lagace and Vizzini (1988).

Compression testing of MMCs has been used to obtain information, not only about damage development and failure mechanisms, but also concerning residual stresses in the matrix. Differences in yielding behaviour under tensile and compressive loading can be related to these residual stress levels. Some of the experimental aspects of such testing have been described by Kennedy (1989). Further information can be obtained by load reversal tests, in which a specimen is plastically deformed in tension and then subjected to compressive loading (or vice versa). MMCs often exhibit a pronounced *Bauschinger effect* (easier yielding after prior reversed plastic flow). Interest has centred mainly on discontinuously reinforced material. Interpretation of experimental data requires some care – see, for example, Taya *et al.* (1990).

Under some circumstances, failure may occur under *transverse compression*, involving shear on planes parallel to the fibre axis. This is expected when a shear stress of the τ_{32} type (see Fig. 8.16) reaches a critical value – which should be broadly similar in magnitude to τ_{12u}. It follows that this type of failure is expected at an applied normal compressive stress of about $2\tau_{12u}$, on planes inclined at 45° to the loading direction and parallel to the fibre axis. Experimental measurements are broadly consistent with this, although there is sometimes a dependence on interfacial bond strength; for example, if a high bond strength leads to failure by shear yielding within a polymeric matrix, then this usually occurs less readily under compression than in tension.

8.2 Failure of laminae under off-axis loads

Failure of laminae subjected to arbitrary (in-plane) stress states can be understood in terms of the three failure mechanisms (with defined values of σ_{1u}, σ_{2u} and τ_{12u}) shown in Fig. 8.1. (It is assumed that large compres-

sive loads and through-thickness stresses are not present.) A number of **failure criteria** have been proposed. The main issue is whether or not the critical stress to trigger one mechanism is affected by the stresses tending to cause the others – i.e. whether there is any *interaction* between the modes of failure.

8.2.1 Maximum stress criterion

In the simple maximum stress criterion, it is assumed that failure occurs when a stress parallel or normal to the fibre axis reaches the appropriate critical value, that is when one of the following is satisfied

$$\sigma_1 \geq \sigma_{1u} \tag{8.7}$$

$$\sigma_2 \geq \sigma_{2u}$$

$$\tau_{12} \geq \tau_{12u}$$

For any stress system (σ_x, σ_y and τ_{xy}) applied to the lamina, evaluation of these stresses can be carried out using Eqn (5.12)

$$\begin{bmatrix} \sigma_1 \\ \sigma_2 \\ \tau_{12} \end{bmatrix} = [T] \begin{bmatrix} \sigma_x \\ \sigma_y \\ \tau_{xy} \end{bmatrix} \tag{[(5.12)]}$$

in which $[T]$ is given by Eqn (5.13)

$$[T] = \begin{bmatrix} c^2 & s^2 & 2cs \\ s^2 & c^2 & -2cs \\ -cs & cs & c^2 - s^2 \end{bmatrix} \tag{[(5.13)]}$$

$$(c = \cos\phi, s = \sin\phi)$$

Monitoring of σ_1, σ_2 and τ_{12} as the applied stress is increased, allows the onset of failure to be identified as the point when one of the inequalities in Eqn (8.7) is satisfied. Noting the form of $[T]$, and considering applied uniaxial tension, the magnitude of σ_x necessary to cause failure can be plotted as a function of loading angle ϕ between stress axis and fibre axis, for each of the three failure modes.

$$\sigma_{xu} = \frac{\sigma_{1u}}{\cos^2\phi} \tag{8.8}$$

$$\sigma_{xu} = \frac{\sigma_{2u}}{\sin^2\phi} \tag{8.9}$$

$$\sigma_{xu} = \frac{\tau_{12u}}{\sin\phi\cos\phi} \tag{8.10}$$

The three curves are plotted in Fig. 8.22, using values of σ_{1u}, σ_{2u} and τ_{12u} appropriate for a polyester/50% glass lamina (see Table 8.2). The solid line indicates the predicted variation of the failure stress as ϕ is increased, according to the maximum stress criterion. Typically, axial failure is expected only for very small loading angles, but the predicted transition from shear to transverse failure may occur anywhere between 20° and 50°, depending on the exact values of τ_{12u} and σ_{2u}.

8.2.2 Other failure criteria

Various other attempts have been made to predict the failure of long-fibre composites under combined stresses, particularly for the plane stress conditions applicable to individual plies in a laminate. A comprehensive review of the approaches adopted has been published by Rowlands (1985). Most treatments are based on adaptations of yield criteria developed for metals. The most common yield criteria are those of Tresca and von Mises. The Tresca criterion corresponds to yield occurring when a critical value of the maximum shear stress is reached. This may be written as

$$[(\sigma_p - \sigma_q)^2 - \sigma_Y^2][(\sigma_q - \sigma_r)^2 - \sigma_Y^2][(\sigma_r - \sigma_p)^2 - \sigma_Y^2] = 0 \qquad (8.11)$$

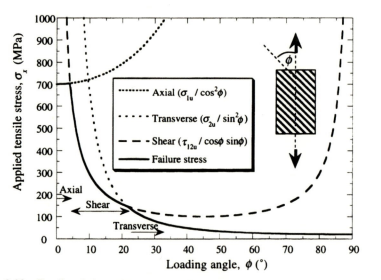

Fig. 8.22   Predicted dependence on loading angle ϕ of the applied stress for the onset of different failure modes for a polyester/50% glass lamina, according to the maximum stress criterion.

where σ_p, σ_q and σ_r are the principal stresses and σ_Y is the yield stress under uniaxial loading. In plane stress ($\sigma_r = 0$) Eqn (8.11) reduces to

$$\sigma_p - \sigma_q = \sigma_Y \tag{8.12}$$

The von Mises criterion corresponds to yield occurring when the distortional (shape-changing) strain energy stored in the material reaches a critical value. This may be expressed

$$(\sigma_p - \sigma_q)^2 + (\sigma_q - \sigma_r)^2 + (\sigma_r - \sigma_p)^2 = 2\sigma_Y^2 \tag{8.13}$$

Under plane stress conditions, this becomes

$$\left(\frac{\sigma_p}{\sigma_Y}\right)^2 - \frac{\sigma_p \sigma_q}{\sigma_Y^2} + \left(\frac{\sigma_q}{\sigma_Y}\right)^2 = 1 \tag{8.14}$$

Adaptation of these criteria to describe failure of composites must take account of the inherent anisotropy of a fibre composite and of the differences between the mechanisms of metal yielding and of composite failure. The effect of anisotropy is also relevant to metals, since these are often highly textured and hence anisotropic in properties. The effect of anisotropy on yielding behaviour was treated in some depth by Hill (1950), who derived a modified von Mises yield criterion for metals with orthotropic symmetry (having three orthogonal planes of symmetry). Under plane stress, this criterion may be written

$$\left(\frac{\sigma_1}{\sigma_{1Y}}\right)^2 + \left(\frac{\sigma_2}{\sigma_{2Y}}\right)^2 - \frac{\sigma_1 \sigma_2}{\sigma_{1Y}^2} - \frac{\sigma_1 \sigma_2}{\sigma_{2Y}^2} + \frac{\sigma_1 \sigma_2}{\sigma_{3Y}^2} + \left(\frac{\tau_{12}}{\tau_{12Y}}\right)^2 = 1 \tag{8.15}$$

where σ_1, σ_2 and τ_{12} are the imposed stresses, referred to the orthogonal directions in the plane, and the material properties σ_{1Y}, σ_{2Y}, σ_{3Y} and τ_{12Y} are the measured yield stresses, in tension and shear, when each is applied in isolation.

It has been common to adapt this criterion to failure of unidirectional composites by replacing the yield stresses by the appropriate measured failure stresses. Since such a composite is transversely isotropic, the failure stresses in the 2- and 3-directions are equal, so that the condition reduces to

$$\left(\frac{\sigma_1}{\sigma_{1u}}\right)^2 + \left(\frac{\sigma_2}{\sigma_{2u}}\right)^2 - \frac{\sigma_1 \sigma_2}{\sigma_{1u}^2} + \left(\frac{\tau_{12}}{\tau_{12u}}\right)^2 = 1 \tag{8.16}$$

This formulation was first proposed by Azzi and Tsai (1965) and Eqn (8.16), commonly known as the **Tsai–Hill criterion**, is widely quoted in composite textbooks and is often used in laminate analysis programs (see

§8.3). It defines an envelope in stress space: if the stress state (σ_1, σ_2 and τ_{12}) lies outside of this envelope, i.e. if the sum of the terms on the left-hand side is equal to or greater than unity, then failure is predicted. The failure mechanism is not specifically identified, although inspection of the relative magnitudes of the terms in Eqn (8.16) gives an indication of the likely contribution of the three modes. Care must be taken that the normal stresses are tensile, since the appropriate failure values may be different in compression (particularly σ_{2u}) – see §8.1.4.

The values of σ_1, σ_2 and τ_{12} can be obtained, for a given angle ϕ between fibre axis and applied stress axis, from Eqn (5.12). For a single applied tensile stress σ_ϕ ($= \sigma_x$ at an angle ϕ to the fibre axis) Eqn (8.16) can be written as

$$\sigma_\phi = \left[\frac{\cos^2\phi(\cos^2\phi - \sin^2\phi)}{\sigma_{1u}^2} + \frac{\sin^4\phi}{\sigma_{2u}^2} + \frac{\cos^2\phi\sin^2\phi}{\tau_{12u}^2} \right]^{-1/2} \tag{8.17}$$

and this expression gives the applied stress at which failure is predicted, as a function of loading angle ϕ.

8.2.3 Experimental data for single laminae

In Fig. 8.23 the predictions of the Tsai–Hill and maximum stress criteria are shown, obtained using the same critical stress levels σ_{1u}, σ_{2u} and τ_{12u}. These critical values were fitted to experimental data for epoxy/50% carbon laminae (obtained by cutting flat rectangular coupons at different angles to the fibre direction and testing them in tension). The experimental data fit rather better to the Tsai–Hill curve, notably for the specimen at $\phi = 30°$ – which was observed to fail by both shear and transverse tension.

It is appropriate at this point to note that there are certain difficulties in carrying out tests designed to obtain critical stress data for different failure modes. For example, there are several complications with the basic off-axis tensile test referred to above. One of these is illustrated by Fig. 8.24. This shows how tensile–shear interactions (see §5.2.2 and Fig. 5.6) can cause distortion of the lamina under load, such that the constraint imposed by the gripping system introduces complex, non-uniform stresses (Pagano and Halpin 1968). Furthermore, specimens can be difficult to grip without slippage, edge effects may be important (see §8.3.3) and there is no freedom to select σ_1, σ_2 and τ_{12} independently so as to discriminate more effectively between failure criteria.

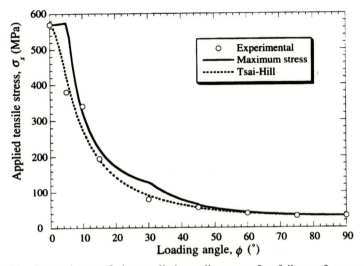

Fig. 8.23 Dependence of the applied tensile stress for failure of epoxy/50% carbon laminae on loading angle. Experimental data (Sinclair and Chamis 1979) are shown, together with predictions from the maximum stress and Tsai–Hill criteria.

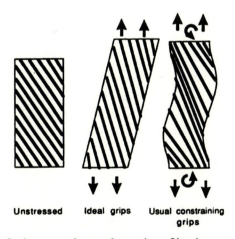

Fig. 8.24 Effect of grip constraint on the testing of laminae under off-axis loads, under conditions such that tensile–shear interaction strains are generated.

A more versatile testing procedure is to produce tubular specimens and test them under various combinations of applied stress. The performance of various types of composite tube and pipe (particularly under internal pressure) is of commercial importance and this topic is covered in §8.4. A simple hoop-wound arrangement, with facilities for applying tension and/ or torsion, is illustrated in Fig. 8.25. Any combination of σ_2 and τ_{12} can be applied, with zero hoop stress (σ_1). For a thin-walled tube of diameter d and wall thickness t ($\ll d$), the shear stress in the wall is obtained from the applied torque T (in N m) by the expression

$$\tau_{12} = \frac{2T}{\pi d^2 t} \tag{8.18}$$

Using a grip section with a diameter larger than the tube under test, as shown in Fig. 8.25, facilitates the application of a large torque, with less danger of slippage in the grips.

Data are shown in Fig. 8.26 giving various combinations of σ_2 and τ_{12} which produced failure in epoxy/65% glass hoop-wound tubes. The failure envelopes for the two criteria are simple and give very different predictions for this case. The maximum stress theory gives the two lines $\tau_{12} = \tau_{12u}$ and $\sigma_2 = \sigma_{2u}$ as limits and the Tsai–Hill condition, Eqn (8.16), reduces to

$$\left(\frac{\sigma_2}{\sigma_{2u}}\right)^2 + \left(\frac{\tau_{12}}{\tau_{12u}}\right)^2 = 1 \tag{8.19}$$

It is clear from the data in Fig. 8.26 that the Tsai–Hill criterion gives a more reliable prediction.

Although the Tsai–Hill criterion is useful, it is not based on rigorous theory and it cannot be applied to all situations. Some of its limitations were highlighted by Tsai and Wu (1971), who proposed a more complex criterion involving both tensile and compressive failure stresses and an interaction parameter which must be measured under biaxial loading.

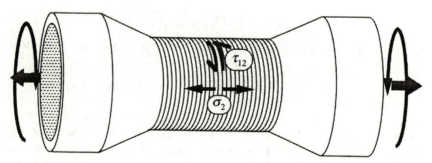

Fig. 8.25 Schematic illustration of how a hoop-wound tube is subjected to simultaneous torsion and tension to investigate failure mechanisms and criteria.

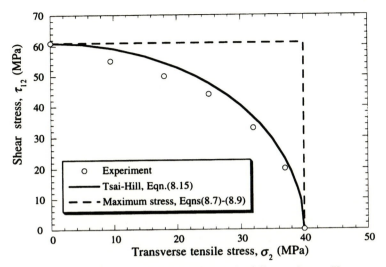

Fig. 8.26 Comparison between experimental failure data (Knappe and Schneider 1973), obtained by combined tension/torsion testing of hoop-wound epoxy/65% glass-fibre composite tubes, and predictions from the maximum stress and Tsai–Hill criteria.

Even with such modifications, however, the interactions between different failure modes are being fitted to an assumed equation, rather than being modelled on the basis of micromechanics. In fact, as pointed out by Hart-Smith (1994), much experimental data can be explained by establishing the complete stress envelope for each possible failure mode, without introducing any interactions between the modes. While this requires extensive experimental data for each system of interest, it may be the best approach when reliable prediction of failure behaviour is essential.

8.3 Strength of laminates

The strength of laminates can be predicted by an extension of the preceding sections, utilising the procedures set out in Chapter 5 for determination of the stresses in the component laminae. For example, Fig. 5.12 showed how the stresses within a crossply laminate under uniaxial tension vary with loading angle. Once these stresses are known (in terms of the applied load), an appropriate failure criterion can be applied and the onset and nature of the failure predicted. However, failure of an individual ply within a laminate does not necessarily mean that the component is no longer usable, as other plies may be capable of withstanding

considerably greater loads without catastrophic failure. Analysis of the
behaviour beyond the initial, fully elastic stage is complicated by uncer-
tainties as to the degree to which the damaged plies continue to bear some
load. Nevertheless, useful calculations can be made in this regime
(although the major interest may be in the avoidance of *any* damage to
the component).

8.3.1 Tensile cracking

Consider first a crossply (0/90) laminate being loaded in tension along
one of the fibre directions. The stresses acting in each ply are shown
schematically in Fig. 5.13, for an epoxy/50% glass composite. Only transverse or axial tensile failure is possible in either ply, since no shear stresses
act on the planes parallel to the fibre directions. The sequence of failure
events when a laminate is loaded progressively in this way is illustrated
schematically in Fig. 8.27. From the data for ϵ_{1u} and ϵ_{2u} shown in Table

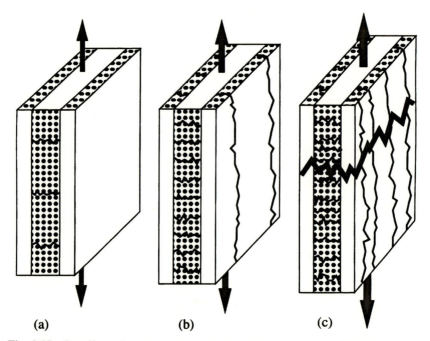

(a)   (b)   (c)

Fig. 8.27   Loading of a crossply laminate parallel to one of the fibre directions.
(a) Cracking of transverse plies as σ_2 reaches σ_{2u}, (b) onset of cracking parallel to
the fibres in axial plies as σ_2 (from inhibition of Poisson contraction) reaches σ_{2u}
and (c) final failure as σ_1 in axial plies reaches σ_{1u}.

8.2, it is clear that transverse plies fail first, as shown in Fig. 8.27(a), despite the fact that they are less highly stressed than axial plies.

Since most of the load is borne by the axial plies (about 85% for the example shown in Fig. 5.13), cracking of the transverse plies does not greatly increase the σ_1 stress and the axial plies usually remain undamaged at this point. As the applied stress increases, the next type of damage to occur is often cracking parallel to the fibres in the axial plies – see Fig. 8.27(b). This is caused by the tensile σ_2 stress, arising from the resistance of the transverse plies to the lateral Poisson contraction of the axial plies. Figures 5.12 and 5.13 show that these stresses are about 7% of the applied stress.

For purposes of comparing σ_1 and σ_2 in the axial plies, a limiting case is to assume that cracking of the transverse plies has made them stress-free in the loading direction, so that σ_1 in the axial plies reaches twice the applied stress. (All of the applied load is taken by the axial ply, which comprises one-half the thickness of the specimen.) Under these circumstances, σ_1 is almost 30 times larger than σ_2 within the axial plies. However, for the three polymer composites listed in Table 8.2, the ratio $(\sigma_{1u}/\sigma_{2u})$ ranges from about 30 to 60. Hence, transverse cracks in the axial plies are frequently observed before the final failure depicted in Fig. 8.27(c). (This is less likely with metallic matrices, for which the ratio σ_{1u}/σ_{2u} is usually smaller and σ_2 stresses are lower because of the lower Poisson's ratio of the matrix: similar arguments apply to ceramic-based laminates.)

Although the laminate depicted in Fig. 8.27(b) has not completely fractured, there is extensive microdamage. The network of cracks is such that the laminate is susceptible to the passage of gases and liquids through the walls, a point of particular relevance for pressure vessels. In addition to concern about leakage, the ingress of certain fluids may hasten final failure – see, for example, the effect of dilute acid in Fig. 8.10.

A typical stress–strain curve for testing a crossply laminate is shown in Fig. 8.28. Also shown on this plot is the output from acoustic emission equipment. These measurements are made with a piezo-electric transducer attached to the specimen surface, which picks up elastic waves triggered by the nucleation and growth of cracks. The onset of transverse cracking is manifest both as an acoustic signal and as a 'knee' in the stress–strain curve, caused by a reduction in the specimen stiffness as the transverse plies are unloaded. This reduction in stiffness is most marked at the beginning, but continues over an appreciable range of strain, as the cracks in the transverse ply become more closely spaced.

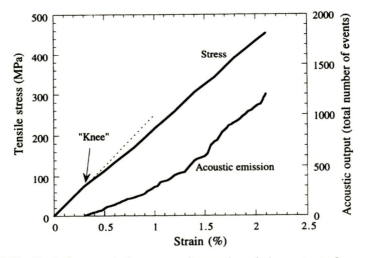

Fig. 8.28   Typical stress–strain curve and acoustic emission output of a crossply
laminate tested in uniaxial tension.

Eventually, however, the axial plies carry virtually all of the load and the
gradient of the plot becomes constant again. There are many factors
which determine the details of how the transverse cracks develop.
Discussions of some of these points can be found in Parvizi and Bailey
(1978), Aveston and Kelly (1980) and McCartney (1992).

8.3.2 *Interlaminar stresses*

Interlaminar stresses are also a source of damage in stressed laminates. It
was pointed out in §5.4 that through-thickness coupling stresses cause
distortions and that these are reduced using 'balanced' stacking
sequences. There are also interlaminar shear stresses operating to transfer
load between laminae and these may give rise to interlaminar cracking. A
common source of damage is the τ_{xz} shear stresses, which arise from
rotation of the individual laminae, as illustrated in Fig. 8.29 for an
angle-ply laminate. This rotation was described in §5.2, during coverage
of the tensile–shear interaction. For an applied tensile stress, the shear
strain resulting from the rotation is proportional to the interaction compliance S_{16}. It follows from Fig. 5.6 that, for ply angles (i.e. loading
angles in Fig. 5.6) below 60°, the fibres tend to rotate towards the stress
axis, while for larger angles than this they tend to rotate so as to lie
normal to it. This is reflected in the data presented in Fig. 8.30, which

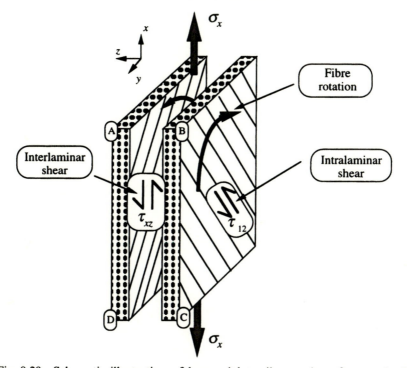

Fig. 8.29 Schematic illustration of how axial tensile stressing of an angle-ply laminate generates a tendency towards rotation of the component plies, giving rise to interlaminar shear stresses.

shows the result of computations carried out by Pipes and Pagano (1970) giving the magnitude of τ_{xz} as a function of ply angle ϕ for an angle-ply epoxy/carbon laminate. The nature of the damage induced by these stresses is illustrated by Fig. 8.31, which is a micrograph of a lateral surface of the type marked ABCD in Fig. 8.29. The matrix shear is evident from the displacement of polishing marks across XX, while the tendency for out-of-plane movement to occur is apparent from the difference in focus on either side of the crack.

8.3.3 Edge effects

The significance of the interlaminar shear failure represented in Figs. 8.29–8.31 depends on the width of the specimen. This type of damage is initiated on the lateral edge surface, shown as ABCD in Fig. 8.29, and can play a dominant role in the overall failure of narrow specimens. This

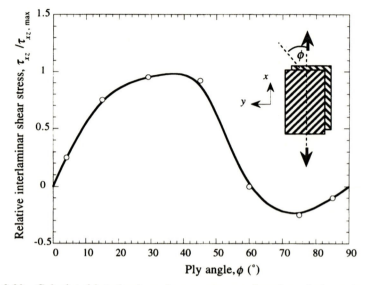

Fig. 8.30   Calculated interlaminar shear stress, as a function of ply angle ϕ, for an angle-ply carbon fibre/epoxy laminate (Pipes and Pagano 1970).

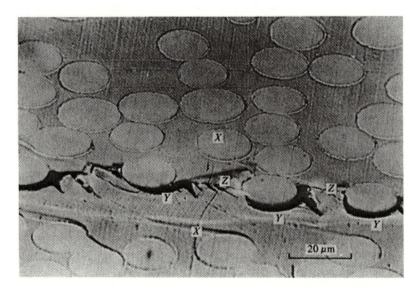

Fig. 8.31   Optical micrograph of an interlaminar crack on the edge face (ABCD in Fig. 8.29) of a glass fibre/polyester resin angle-ply laminate (Jones 1981).

effect is illustrated by the data in Fig. 8.32, which show the dependence on specimen width of the stresses at which transverse cracking and final fracture occur for a $\pm 50°$ laminate. If the specimen is very narrow, interlaminar shear cracking becomes the dominant failure mode and it can occur at such a low applied stress that transverse cracking is completely suppressed. These effects are most pronounced for the low ply angles at which τ_{xz} is predicted to be higher. In addition, the magnitude of the fracture stress increases as ϕ is reduced and σ_1 increases relative to σ_2 and τ_{12}. These two trends are apparent in the data shown in Fig. 8.33. Thus, care should be taken in specifying the specimen width for such tests. For certain ply angles, only very wide specimens allow the effect of interlaminar shear to be eliminated when interpreting the results. This is one reason for the interest in testing of tubes, treated in §8.4.

8.4 Failure of tubes under internal pressure

In addition to the industrial significance of the failure behaviour of composite tubes under internal pressure, this mode of testing is a convenient

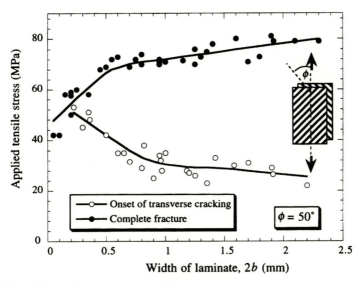

Fig. 8.32 Effect of test specimen width on tensile stress for transverse cracking and complete fracture in angle-ply laminates ($\phi = 50°$) of polyester/50% glass fibre (Jones 1981).

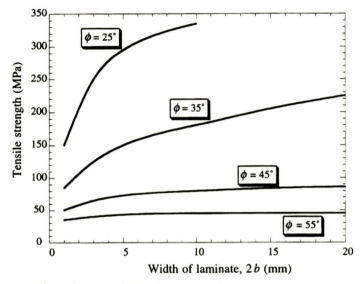

Fig. 8.33   Effect of test specimen width on tensile fracture strength of polyester/
50% glass fibre angle-ply laminates, for various ply angles (Jones 1981).

method of generating selected stress states in a laminate and avoiding the complications of edge effects. Internal pressure P acts on the cylindrical surfaces of a tube to cause a hoop stress σ_H in the tube wall and on the sealed ends to generate an axial stress σ_A. Provided the wall thickness t is much less than the tube radius r, these stresses are given by

$$\sigma_H = \frac{Pr}{t} \tag{8.20}$$

$$\sigma_A = \frac{Pr}{2t} = \frac{\sigma_H}{2} \tag{8.21}$$

Two modes of testing are commonly used, as depicted in Fig. 8.34. When the pressure is contained with rubber 'O' rings, on which the tube is free to slide, the axial stresses are eliminated, so that $\sigma_A = 0$. When the ends of the tube are sealed, the axial stresses are given by Eqn (8.21). When testing is to be continued beyond initial cracking to final failure, a rubber lining is used to eliminate leakage of pressurising fluid.

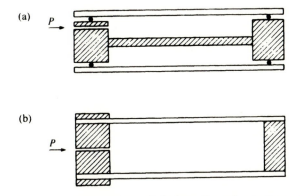

Fig. 8.34 Schematic representation of designs for tube testing with (a) pure hoop ($\sigma_A = 0$) and (b) hoop plus axial ($\sigma_A = \sigma_H/2$) loading.

8.4.1 Pure hoop loading

Angle-ply tubes can be manufactured using the filament winding methods outlined in §11.1.1. When tested under pure hoop loading, using the set-up of Fig. 8.34(a), the tube wall is subjected to the same loading conditions as for testing a flat laminate under uniaxial tension. The stresses in the component plies can be calculated using the methods of Chapter 5. The failure stresses can be predicted using a failure criterion and measured values of σ_{1u}, σ_{12u} and τ_{2u}. The results shown in Fig. 8.35, which were obtained using the maximum stress criterion, demonstrate how the stresses in the plies, and the failure stress, vary with ply angle ϕ for polyester/50% glass laminates. (The values used for σ_{1u}, σ_{2u} and τ_{12u} are those in Table 8.2.)

The appearance of the failure stress plot (Fig. 8.35(b)) is broadly, but not exactly, the same as that for a single lamina shown in Fig. 8.22: the differences are due to the constraint imposed on each ply by the presence of the other ply. Simple estimates of the hoop stress at failure can nevertheless be made by treating a single ply in isolation, so that Eqn (5.12) can be used in conjunction with, for example, the Tsai–Hill criterion to give Eqn (8.16): a similar procedure may also be employed when both hoop and axial stresses are present.

Experimental studies of the stress–strain behaviour of such tubes lead to data of the form shown in Fig. 8.36. The curves depart from initial linearity as cracking proceeds (and possibly also as a result of some

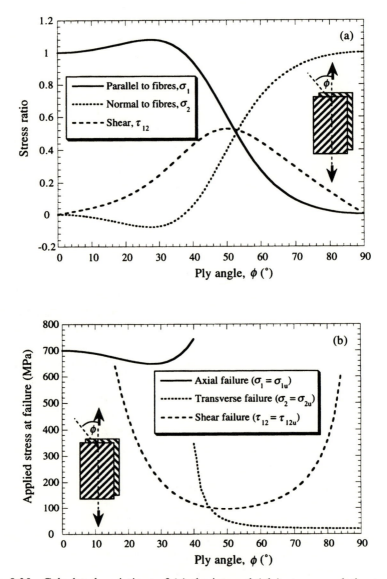

Fig. 8.35 Calculated variations of (a) the internal (ply) stresses, relative to an applied tensile stress and (b) the applied stress needed to cause different types of failure, as a function of the ply angle ϕ for an angle-ply laminate of polyester/50% glass fibre.

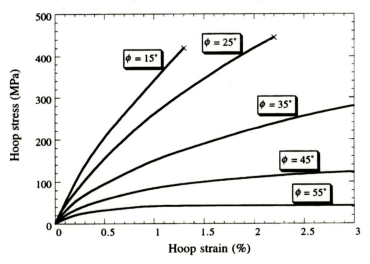

Fig. 8.36 Hoop stress against hoop strain curves of polyester/50% glass fibre angle-ply laminated tubes tested in pure hoop loading (Spencer and Hull 1978).

viscoelastic flow of the resin matrix). The onset of non-linearity occurs earlier for the higher ply angles, as expected from Fig. 8.35(b). For ϕ greater than about 40°, initial damage is by transverse cracking, but this cannot happen for the lower values since σ_2 becomes negative – see Fig. 8.35(a). The departure from linearity occurs in these cases as a result of shear displacements.

8.4.2 Combined hoop and axial loading

When tubes are tested with sealed ends, as shown in Fig. 8.34(b), the laminate is subject to biaxial tension, with tensile stresses of σ_H in the hoop direction and $\sigma_A = \sigma_H/2$ normal to this in the tube axis direction. Results for this case are shown in Fig. 8.37. The transverse stress σ_2 is now always positive and significant, with the result that transverse cracking is predicted to be the dominant initial failure mode for all ply angles. These predictions may be compared with the experimental stress–strain data shown in Fig. 8.38. As predicted, initial failure occurred by transverse cracking in all cases, causing departure from linearity at low stresses. This was more delayed for the ply angle of

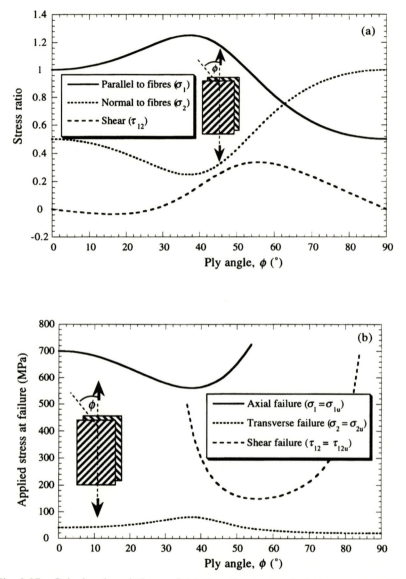

Fig. 8.37 Calculated variations of (a) the ply stresses (relative to an applied tensile stress) and (b) the applied stress needed to cause different types of failure, as a function of the ply angle ϕ for an angle-ply laminate of polyester/50% glass fibre. In this case, there is also a second applied tensile stress, normal to the first and one-half its magnitude, as would be the case for an internally pressurised tube.

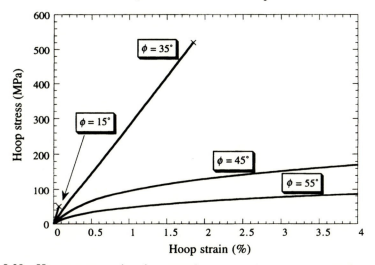

Fig. 8.38 Hoop stress against hoop strain curves of polyester/50% glass fibre angle-ply laminated tubes tested in combined hoop and axial loading (Spencer and Hull 1978).

35°, in agreement with Fig. 8.37(b). Prediction of the point of final fracture is complicated by the unloading of damaged plies, and by the stress concentration effects of cracks. As a final topic in this chapter, a simple approach is outlined to describe the effects of the initial damage on subsequent behaviour.

8.4.3 Netting analysis

In netting analysis, it is assumed that only the fibres bear load, so that $\sigma_2 = \tau_{12} = 0$. Although this is not very realistic, it is broadly valid after considerable intralaminar and interlaminar cracking and shearing have taken place. Analysis of this case cannot be carried out rigorously, but an approximate estimate can be made as follows. For an angle-ply laminate, each lamina is considered independently and the ply angle ϕ taken as the angle between the fibre axis and the x-axis (i.e. the hoop direction). Using the inverse of Eqn (5.12) to relate applied and internal stresses, with $\sigma_x = \sigma_H$, $\sigma_y = \sigma_H/2$ and $\sigma_2 = \tau_{12} = 0$ gives

$$\sigma_H = \sigma_1 \cos^2 \phi$$
$$\frac{\sigma_H}{2} = \sigma_1 \sin^2 \phi$$

(8.22)

There is potential for confusion, because Eqn (5.12) apparently shows that σ_1 is equal to $\sigma_H \cos^2 \phi + (\sigma_H \sin^2 \phi)/2$, which cannot exceed σ_H. This arises because a single ply cannot in fact support the hoop and axial stresses, while maintaining $\sigma_2 = 0$, without the other ply being present to prevent failure in the transverse (2) direction. Accepting that the treatment is simplified, it may be noted that there is a unique value of ϕ for which Eqns (8.22) both apply

$$\phi = \tan^{-1}\left(\frac{1}{\sqrt{2}}\right) \approx 35° \qquad (8.23)$$

This is sometimes called the 'ideal' angle for an internally pressurised tube. If the netting analysis assumptions hold, then the fibres in laminates with ply angles other than 35° must rotate towards this angle. This gives an insight into the behaviour noted in Fig. 8.38, which shows a much greater failure stress for this ply angle: in practice, rotation towards this angle for other laminates requires extensive matrix damage and this is likely to stimulate fibre fracture.

The analysis can also be used to explore the stiffness after initial damage, and the final failure stress, for the $\phi = 35°$ case. Under netting analysis conditions

$$\sigma_1 = \frac{\sigma_H}{\cos^2 \phi} = 1.5\, \sigma_H \qquad (8.24)$$

and, given that only the fibres contribute to the stiffness ($E_1 = f\, E_f$), it follows that, since the axial, hoop and fibre strains are all equal in this special case, the hoop stiffness is given by

$$E_H = \frac{E_f f}{1.5} \qquad (8.25)$$

This has a value of about 25–30 GPa for polyester/50% glass, which is in good agreement with the gradient of the $\phi = 35°$ plot in Fig. 8.38, after the initial portion; this may be compared with the value in the elastic regime of about 90–100 GPa, obtained both experimentally and from laminate analysis. Final failure occurred at $\sigma_H \sim 500$ MPa, indicating a σ_{1u} value of about 750 MPa and hence a fibre strength of about 1.5 GPa. This is about one-half the strength of freshly drawn glass fibres. The appearance of the final fracture site is shown in Fig. 8.39. Estimates of this type can be useful, provided the implications of the netting analysis assumptions are fully appreciated.

Fig. 8.39   Photograph of final failure region of an angle-ply tube tested in biaxial tension. A crack has formed parallel to one set of fibres and fibre fracture has occurred in the other set. (Hull *et al.* 1978).

References and further reading

Adams, D. F. and Dorner, D. R. (1967) Longitudinal shear loading of a unidirectional composite, *J. Comp. Mat.*, 1 4–17

Argon, A. S. (1972) *Fracture of Composites*. Academic: New York

Aveston, J. and Kelly, A. (1980) Tensile first cracking strain and strength of hybrid composites and laminates, *Phil. Trans. Roy. Soc. London*, **A294** 519–34

Azzi, V. D. and Tsai, S. W. (1965) Anisotropic strength of composites, *Expl. Mechanics*, **5** 283–8

Batdorf, S. B. and Gaffarian, R. (1984) Size effect and strength variability of unidirectional composites, *Int. J. Fracture*, **26** 113–23

Cook, J. and Gordon, J. E. (1964) A mechanism for the control of crack propagation in all-brittle systems, *Proc. Roy. Soc. London*, **A282** 508–20

Ewins, P. D. and Potter, R. T. (1980) Some observations on the nature of fibre reinforced plastics and implications for structural design, *Phil. Trans. Roy. Soc. London*, **A294** 507–17

Hart-Smith, L. J. (1994) Should fibrous composite failure modes be interacted or superimposed?, *Composites*, **24** 53–5

He, M. Y., Evans, A. G. and Curtin, W. A. (1993) The ultimate tensile strength of metal and ceramic matrix composites, *Acta Metall.*, **41** 871–8

Hill, R. (1950) *The Mechanical Theory of Plasticity*. Oxford University Press: London.

Hull, D., Legg, M. J. and Spencer, B. (1978) Failure of glass/polyester filament wound pipe, *Composites*, **9** 17–24

Jelf, P. M. and Fleck, N. A. (1992) Compression failure mechanisms in unidirectional composites, *J. Comp. Mat.*, **26** 2706–26

Jones, M. C. L. (1981) PhD thesis, University of Liverpool

Kelly, A. and Macmillan, N. H. (1986) *Strong Solids*. Clarendon: Oxford

Kennedy, J. M. (1989) Tension and compression testing of MMC materials, in *Metal Matrix Composites: Testing, Analysis and Failure Modes. ASTM STP 1032*, W. S. Johnson (ed.) pp. 7–18

Knappe, W. and Schneider, W. (1973) The role of failure criteria in the fracture analysis of fibre–matrix composites, in *Deformation and Fracture of High Polymers*. H. H. Kausch *et al.* (eds.) Plenum: New York pp. 543–56

Lagace, P. A. and Vizzini, A. J. (1988) The sandwich column as a compressive characterisation specimen for thin laminates, in *Composite Materials: Testing and Design. ASTM STP 972*, J. D. Whitcomb (ed.) pp. 148–60

Legg, M. J. (1980), PhD thesis, University of Liverpool

McCartney, L. N. (1992), Theory of stress transfer in a 0° − 90° − 0° cross-ply laminate containing a parallel array of transverse cracks, *J. Mech. Phys. Sol.*, **40** 27–68

Pagano, N. J. and Halpin, J. C. (1968) Influence of end constraints in the testing of anisotropic bodies, *J. Comp. Mats.*, **2** 18–31

Parry, T. V. and Wronski, A. S. (1981) Kinking and tensile compressive and interlaminar shear failure mechanisms in CFRP beams tested in flexure, *J. Mat. Sci.*, **16** 439–50

Parvizi, A. and Bailey, J. E. (1978) On multiple transverse cracking in glass fibre epoxy cross-ply laminates, *J. Mat. Sci.*, **13** 2131–6

Pipes, R. B. and Pagano, N. J. (1970) Interlaminar stresses in composite laminates under axial extension, *J. Comp. Mat.*, **4** 538–48

Prewo, K. M. and Krieder, K. R. (1972) The transverse tensile properties of boron fibre reinforced aluminium matrix composites, *Metall. Trans.*, **3** 2201–11

Rosen, B. W. (1965) Mechanics of composite strengthening, in *Fibre Composite Materials*. ASM: Metals Park, Ohio chapter 3

Rowlands, R. E. (1985) Strength (failure) theories and their experimental correlation, in *Handbook of Composites, Vol. 3 – Failure Mechanics of Composites*. G. C. Sih and A. M. Skudra (eds.) Elsevier: Amsterdam pp. 71–125

Sinclair, J. H. and Chamis, C. C. (1979) Fracture modes in off-axis fibre composites, in *Proc. 34th SPI/RP Tech. Conf., paper 22A*. Soc. of Plastics Industry: New York

Spencer, B. and Hull, D. (1978) Effect of winding angle on the failure of filament wound pipe, *Composites*, **9** 263–71

Taya, M., Lulay, K. E., Wakashima, K. and Lloyd, D. J. (1990) Bauschinger effects in SiCp–6061 Al composite, *Mat. Sci. & Eng.*, **A124** 103–11

Tsai, S. W. and Wu, E. M. (1971) A general theory of strength for anisotropic materials, *J. Comp. Mat.*, **5** 58–80

Watson, M. C. and Clyne, T. W. (1993) Reaction-induced changes in interfacial and macroscopic mechanical properties of SiC monofilament reinforced titanium, *Composites*, **24** 222–8

Wu, E. M. (1974) Phenomenological anisotropic failure criterion, in *Composite Materials Vol. 2 – Mechanics of Composite Materials*. G. P. Sendeckyj (ed.) Academic: New York pp. 353–431

Zweben, C. and Rosen, B. W. (1970) A statistical theory of material strength with application to composite materials, *J. Mech. Phys. Sol.*, **18** 189–206

9

Toughness of composites

The previous chapter covered factors affecting strength, which is related to the stresses at which damage and failure occur in composites. In many situations, the energy absorbed by the material under load is equally important. A tough material is one for which large amounts of energy are required to cause failure. In many loading configurations, such as when a component is struck by a projectile, only a finite amount of energy is available to cause failure. In other cases, such as with loads arising from temperature changes, only a finite degree of strain needs to be accommodated in order for the stresses to become small. In such situations, toughness, rather than strength, is the key property determining whether the material is suitable. In this chapter, a brief outline is given of the basics of fracture mechanics, with particular reference to the energetics of interfacial damage. This is followed by an appraisal of the sources of energy absorption in composites. Finally, slow crack growth in composites is examined for conditions where fast fracture is not energetically favoured.

9.1 Fracture mechanics

The reader is referred to sources such as Gordon (1978), Ashby and Jones (1980) and Ewalds and Wanhill (1984) for introductions to fracture mechanics. In this section, the treatment is abbreviated and oriented towards effects in composites.

9.1.1 Basic concepts

Fracture mechanics has its roots in the work of Inglis (1913), who demonstrated that the stresses near a crack tip in a material under load are often

208

much higher than that in the bulk, σ_∞. He derived the following expression for the stress at the tip of a crack of length c (or $2c$ if internal) and tip radius r

$$\sigma = \sigma_\infty \left(1 + 2\sqrt{\frac{c}{r}}\right) \tag{9.1}$$

Thus, a circular hole ($c = r$) has a **stress concentration factor** of 3. While this is physically reasonable, the case of a sharp crack ($r \to 0$) presents difficulties, in that the stress concentration, according to Eqn (9.1), becomes very large. On this basis most structures should fail, under low applied loads, at the fine surface scratches which are almost inevitably present. This was contrary to engineering experience at the time, since most artefacts, particularly metallic ones, were able to function even when small cracks and scratches were present.

The problem was resolved by the pioneering work of Griffith (1920), who pointed out that a crack cannot propagate unless the energy of the system is thereby decreased. The energy released when a crack advances comes from the associated release of stored elastic strain in the surrounding material (plus any work done by the loading system). If this is insufficient to counterbalance the energy absorbed in the material through the creation of new fracture surfaces and associated internal damage or deformation processes, then the crack cannot advance. In many materials, efficient mechanisms for internal energy absorption are stimulated by the high stresses at a crack tip, so that the energy balance for crack propagation is often unfavourable and they exhibit high toughness (resistance to fracture). This is particularly true for most metals, since the dislocation motion which occurs is highly effective in this regard.

Griffith considered brittle materials, such as glasses, in which energy-absorbing processes are not readily stimulated and the only significant energy penalty of crack propagation comes from the new surface area of the crack faces. He showed that the change in the stored energy of a loaded plate of unit thickness, caused by the introduction of an interior crack of length $2c$, is given by

$$U = -\frac{\sigma^2 \pi c^2}{E} \tag{9.2}$$

where $\sigma(= \sigma_\infty)$ is the applied stress and E is the Young's modulus. The other contribution to the overall energy change is that required to create the new surface area, which is positive and has a value of $4c\gamma$, where γ is the surface energy. The dependence of these two contributions on the length of the crack is shown in Fig. 9.1. Only cracks longer than a critical length, c_*, will grow spontaneously (with reduction in net energy). This critical length is found by differentiating the total energy with respect to crack length and setting the result equal to zero, leading to

$$c_* = \frac{2\gamma E}{\sigma^2 \pi} \qquad (9.3)$$

This approach was extended by Irwin (1948) to encompass tougher materials. The surface energy term 2γ is supplemented by other contributions to the energy absorbed in the vicinity of an advancing crack tip. For a given applied stress and pre-existing crack size, an expression can be obtained from Eqn (9.3) for the *energy release rate*, G, which has units of $J\,m^{-2}$

$$G = \frac{\sigma^2 \pi c}{E} \qquad (9.4)$$

For fracture to occur, this must exceed a critical value, sometimes referred to as the *crack resistance*, R. This critical value represents the total energy absorbed, per unit of crack advance area and is often termed G_c, the *critical energy release rate*, or *fracture energy*. Values of G_c are

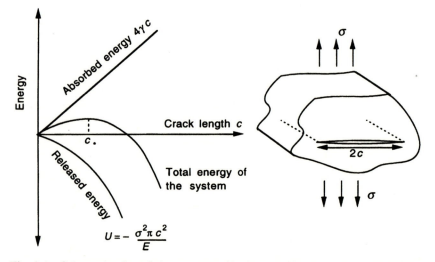

Fig. 9.1 Schematic plot of the two contributions to the energy associated with the presence of a crack in a brittle material, as a function of crack length. A crack of length c_* or larger will grow spontaneously, with a reduction in the total energy.

Table 9.1 *Typical fracture energy and fracture toughness values for various materials. (After Ashby and Jones 1980)*

Material	Fracture energy G_c (kJ m^{-2})	Fracture toughness K_c (MPa $\sqrt{m}$)
Polymers		
epoxy resins	0.1–0.3	0.3–0.5
Nylon 6.6	2–4	3
polypropylene	8	3
Metals		
pure Al	100–1000	100–350
Al alloy	8–30	23–45
mild steel	100	140
Ceramics		
soda glass	0.01	0.7
SiC	0.05	3
concrete	0.03	0.2
Natural materials		
woods (crack ⊥ grain)	8–20	11–13
woods (crack // grain)	0.5–2	0.5–1
bone	0.6–5	2–12
Composites		
fibreglass (glass/epoxy, planar random fibres)	40–100	42–60
Al-based particulate MMC	2–10	15–30
SiC laminate (crack ⊥ layers)	5–8	45–55

fairly easy to obtain experimentally. For example, the work done in a tension or bending test is given by the area under a load–displacement plot and, provided this energy is all permanently absorbed in the specimen, the fracture energy is then found by simply dividing by the sectional area through which failure has occurred. The specimen is commonly pre-notched so as to ensure that crack propagation occurs. Typical values of G_c are given in Table 9.1 for various materials. Tough (soft) metals have fracture energies of $100\,\mathrm{kJ\,m^{-2}}$ or more, whereas a brittle material, such as glass, can have a value as low as $0.01\,\mathrm{kJ\,m^{-2}}$. Rearranging Eqn (9.4), the stress necessary to cause spontaneous fracture in a component with a pre-existing crack of size c ($2c$ if internal) can be written as

$$\sigma_* = \sqrt{\frac{G_c E}{\pi c}} \qquad\qquad (9.5)$$

This approach is particularly useful in practical terms, because attention is diverted from the complex problem of the precise nature of the stress field close to the tip of a crack to a more global approach involving macroscopic quantities which are measurable experimentally. However, there is still interest in the phenomena occurring locally near the crack tip. A useful link is provided between the energy and stress field approaches by the concept of a *stress intensity factor*, K. This parameter, which largely evolved from the work of Irwin in the 1950s, can be expressed as

$$K = \sigma\sqrt{\pi c} \tag{9.6}$$

It therefore encompasses the effects of both the applied load and the pre-existing crack size, with the relative weighting that these two parameters have in determining the value of G, the energy release rate (see Eqn (9.4)). It characterises the severity of the stress field around the crack tip. A critical value can be identified, corresponding to the case where the associated value of G reaches G_c

$$K_c = \sigma_*\sqrt{\pi c} = \sqrt{EG_c} \tag{9.7}$$

This *critical stress intensity factor* is often known as a *fracture toughness*. Values are given in Table 9.1 for various materials. For tough materials, the fracture toughness can exceed $100\,\text{MPa}\,\sqrt{m}$, while a brittle material might typically have a value around $1\,\text{MPa}\,\sqrt{m}$. The usefulness of the stress intensity factor lies largely in the way it can be related to local crack tip features. For example, it can be shown that the size of the *plastic zone* ahead of the crack tip is related to the yield stress of the material by

$$r_Y \approx \frac{1}{2\pi}\left(\frac{K}{\sigma_Y}\right)^2 \tag{9.8}$$

Similarly, the *crack opening displacement*, δ, can be expressed as

$$\delta \approx \left(\frac{K^2}{\sigma_Y E}\right) \tag{9.9}$$

Such parameters are useful when considering how energy-absorbing processes might be stimulated in composite materials, since they allow the scale of features of the crack tip to be related to the scale of the microstructure.

9.1.2 Interfacial fracture and crack deflection

Table 9.1 shows that a composite made from glass fibres and epoxy resin has a fracture energy comparable with those of metals ($G_c \sim 50\,\mathrm{kJ\,m^{-2}}$), even though the constituents are both brittle ($G_c \sim 0.01\text{--}0.1\,\mathrm{kJ\,m^{-2}}$). This high toughness of composites, which is very important in practical terms, is closely linked with interfacial effects. A first step in exploring this is to analyse the conditions under which interfacial debonding, i.e. crack propagation along an interface between two different materials, occurs. For a given loading configuration, the propagation of a crack along an interface between two constituents gives rise to an energy release rate, G_i, in much the same way as for the case when the crack is in a homogeneous material. Also, there is a critical value, G_{ic}, an ***interfacial fracture energy***, which G_i must reach for the crack to propagate.

Values of G_{ic} are not as readily available as G_c values for homogeneous materials. There are several reasons for this. Firstly, the toughness of an interface is sensitive to the way in which the interface was manufactured, rather than being unique to the pair of constituents on either side. A second reason is slightly more complex. Interfacial cracks often propagate under ***mixed mode*** loading conditions. This is in contrast to a crack in a homogeneous material, which will always tend to advance in a direction such that the stress field at the crack tip is purely tensile (mode I). An interfacial crack, however, is constrained to follow a pre-determined path. Depending on the loading configuration, the stress field at the crack tip may include a significant shear stress component acting on the plane of the interface (mode II). In general, the energy expended in debonding the interface is greater when there is a mode II component than for the case of pure mode I loading. This complicates the experimental measurement of G_{ic}. Not only can it be difficult to establish the exact stress field at the crack tip, but it may vary with position in the specimen, particularly for fibre/matrix interfaces. The situation is further complicated if any residual stresses (e.g. see §10.1.1) are present.

The proportion of opening and shearing modes at the crack tip is often characterised by means of the ***phase angle***, ψ (psi). This is defined in terms of the mode I and mode II stress intensity factors, shown schematically in Fig. 9.2.

$$\psi = \tan^{-1}\left(\frac{K_{II}}{K_I}\right) \tag{9.10}$$

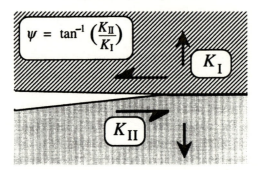

Fig. 9.2   Schematic depiction of the stress field at an interfacial crack. The phase angle ψ, used to characterise the crack tip loading mode, is defined in terms of the stress intensity factors, K_I and K_{II}, for mode I and mode II loading. The value of ψ can vary from $0°$ (pure mode I, crack opening) to $90°$ (pure mode II, shearing).

The value of ψ is $0°$ for pure opening ($K_{II} = 0$) and $90°$ for pure shear ($K_I = 0$). The phase angle can be established for various loading arrangements, although often this calculation is not a simple one (Rice 1988, Evans and Hutchinson 1989, Wang and Suo 1990). Furthermore, ψ is likely to vary with position when the interface is non-planar. The limited experimental data currently available suggest that the dependence of G_{ic} on ψ may be quite significant, depending on the type of system. Examples are shown in Fig. 9.3 for planar interfaces in the form of thin interlayers between thicker substrates. These data were obtained using different tests and covering a wide range of opening and shearing mode contributions. Increases in G_{ic} appear to result when the shearing mode component is increased. This may be due, at least partly, to more frictional work being done immediately behind the crack tip as asperities on the crack flanks slide over each other. The dependence on ψ appears to be stronger when at least one of the constituents can deform plastically (see Fig. 9.3(b)).

One of the main reasons for interest in interfacial toughness concerns *crack deflection*. For a composite to have a high toughness, a crack passing through the matrix must be repeatedly deflected at fibre/matrix interfaces, at least for materials based on polymer resins or ceramics (see §9.2). Early work by Cook and Gordon (1964) on conditions for crack deflection was focused on the stress field ahead of a crack tip. It was pointed out that, when a crack approaches a fibre in a composite loaded parallel to the fibre axis, there is a transverse stress ahead of the crack tip, tending to open up the interface and hence to 'blunt' and deflect the crack from entering the fibre (§8.1.1). Since the peak value of this transverse

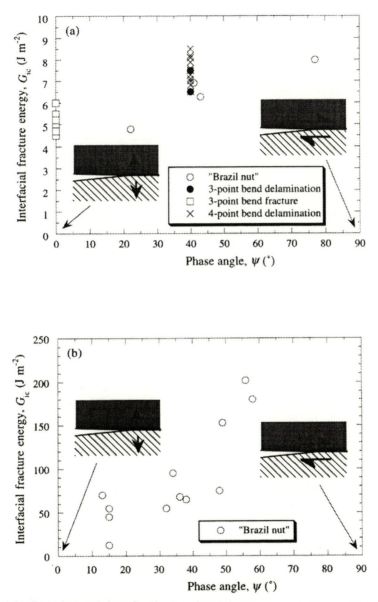

Fig. 9.3 Experimental data for the interfacial fracture energy G_{ic} as a function of the phase angle ψ which characterises the stress field at the crack tip. (a) Thin ($\sim$ 5–10 μm) graphitic layers between SiC sheets (Phillipps *et al.* 1993), and (b) thin ($\sim$ 100–400 μm) epoxy layers between aluminium blocks (Akisanya and Fleck 1992). The legend refers to the type of test used to obtain the data.

stress is always about 20% of the maximum axial stress, Cook and Gordon proposed that the interface debonds if its strength is less than about one-fifth that of the matrix. However, the 'strength' of an interface is not well defined and is sensitive to the presence and distribution of flaws and to the method of loading. In view of the success of the Griffith treatment, it is clearly preferable to establish a criterion based on the energetics of crack propagation.

Energy-based crack deflection criteria have been proposed by Kendall (1975) and by He and Hutchinson (1989). Kendall considered two blocks bonded together and loaded in tension parallel to the interface, one of the blocks having a crack approaching the interface – see Fig. 9.4. He estimated the applied loads under which a crack would penetrate the other block or deflect along the interface, assuming that the crack requiring the lower load would predominate. This produced the following criterion for deflection

$$\frac{G_{ic}}{G_{fc}} \leqslant \left(\frac{h_m E_m + h_f E_f}{h_f E_f}\right)\left[\frac{1}{4\pi(1-\nu^2)}\right] \tag{9.11}$$

where G_{ic} and G_{fc} are the fracture energies of interface and uncracked block (fibre), h_m and h_f are thicknesses of cracked (matrix) and uncracked (fibre) blocks, E_m and E_f are the corresponding Young's

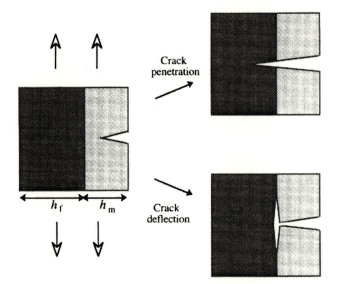

Fig. 9.4 Geometry of crack deflection at an interface (after Kendall (1975)).

moduli and ν is the Poisson's ratio (taken as equal in both constituents). If h_m and h_f are taken as equal, corresponding approximately to a crack passing through the matrix between unbroken fibres in a typical composite, then this critical ratio is around 20% for $E_m \sim E_f$, falling to about 10% for $E_m \ll E_f$.

He and Hutchinson's analysis is based on considering whether the penetrating or deflecting crack gives a greater net release of energy. It is more complex than Kendall's model, but also yields a critical fracture energy ratio of around 20% for the case of $E_m \sim E_f$. For $E_m \ll E_f$, however, an increase (rather than a decrease) is predicted in this ratio. There are few experimental data available to validate either of these models. Kendall's experimental work with rubbers did seem consistent with his predictions, but few systematic measurements have been performed with systems of practical interest. It is, however, clear that the interfacial fracture energy must be appreciably lower than that of the reinforcement if matrix cracks are to be consistently deflected. Since (ceramic) fibres tend to have low fracture energies (see Table 9.1 and §9.2.2), this means that interfaces of very low toughness are often required if crack deflection is essential. For ceramic matrix composites, in particular, the retention of low interfacial toughness, through processing stages which tend to promote sintering and chemical reactions, represents a major technological challenge.

9.2 Contributions to work of fracture

A relatively high toughness is essential for most engineering materials. One of the advantages of composite materials is that there is often scope for the promotion of energy-absorbing mechanisms in the material. It is important to understand how these mechanisms are controlled.

9.2.1 Matrix deformation

Work of fracture data for some typical matrices are given in Table 9.1. Most metallic matrices have a high toughness, mainly as a result of the extensive dislocation movement which occurs near a crack. Polymers, particularly thermosets, and ceramics have low fracture energies.

The extent of matrix deformation during composite fracture may differ appreciably from that in the same material when unreinforced. The main effect is one of increased **constraint**, so that the matrix is unable to deform freely because it is surrounded by stiff and strong fibres. This may be partly a result of **load transfer** (§1.3), which reduces the magnitude of the matrix stresses. Of greater significance, however, is the tendency for **triaxial stress states** to be set up which inhibit plastic flow of the matrix. For example, when a region of matrix extends plastically, with associated lateral contraction, this contraction may be opposed by a surrounding rigid cluster of fibres. This sets up transverse tensile stresses which reduce the deviatoric (shape-changing) component of the matrix stress state. This in turn inhibits plastic flow, but may encourage cavitation and fracture.

This type of effect accounts for the lower fracture energy shown in Table 9.1 for the Al-based MMC, when compared with unreinforced Al. This loss of toughness can be minimised by eliminating reinforcement clusters and other inhomogeneities such as pores, debonded interfaces and cracked particles or fibres. Processing improvements for MMCs are currently aimed in this direction. It follows that a high interfacial strength is desirable for MMCs and in most cases this is quite readily achievable (§7.3). Only for relatively low-toughness metals (e.g. zinc) reinforced with long fibres is there any interest in a low interfacial toughness, since in these cases toughening by fibre pull-out (§9.2.4) is preferable to retaining as much of the toughness of the matrix as possible (Vescera *et al.* 1991).

The constraint effects outlined above also operate with non-metallic matrices. In these cases, however, there is in any event limited scope for energy absorption in the matrix and other mechanisms are often of greater significance. However, considerable improvements in the toughness of PMCs have been achieved by increasing the toughness of the matrix, which can alter the micromechanisms of damage and increase the associated energy absorption.

9.2.2 Fibre fracture

Depending on the fibre architecture (Chapter 3) and loading configuration, component failure usually involves the fracture of fibres. The contribution this makes to the fracture energy of the material is small for most fibres. Typical fracture energies for fibres of glass, carbon and SiC are only a few tens of $J\,m^{-2}$. Some polymeric fibres are not completely brittle and undergo appreciable plastic deformation. An example of this can be seen in Fig. 2.7(d), which shows a fractured Kevlar™ fibre. Most natural fibres are in a similar category and fracture of the cellulose fibres makes a significant contribution to the fracture energy of wood (across the grain). In such cases, fibre fracture contributes up to a few $kJ\,m^{-2}$ to the overall fracture energy. Metallic fibres can in principle make even larger contributions. Thus, even at a low volume content, steel rods in reinforced concrete raise the toughness substantially, as well as enhancing the tensile strength. Nevertheless, for most composites the fibres themselves make little or no direct contribution to the overall toughness.

9.2.3 Interfacial debonding

Some interfacial debonding usually occurs during the fracture of a composite. If a crack is propagating normal to the direction of fibre alignment, debonding occurs provided the crack is deflected on reaching the interface. Debonding can also be stimulated under transverse or shear loading conditions. Provided the value of the interfacial fracture energy, G_{ic}, is known (see §9.1.2), an estimate can be made of the contribution from debonding to the overall fracture energy.

The basis for such a calculation is shown in Fig. 9.5 for aligned short fibres. A simple shear lag approach is used. Provided the fibre aspect ratio, $s\ (= L/r)$, is less than the critical value, $s_*\ (= \sigma_{f^*}/2\tau_{i^*})$, see §6.1.5, all of the fibres intersected by the crack debond and are

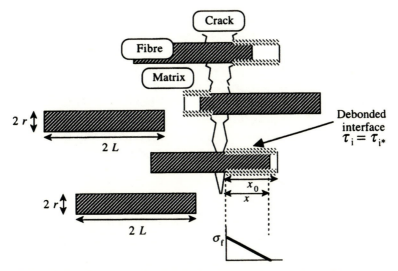

Fig. 9.5 Schematic of a crack passing through an aligned short-fibre composite, showing interfacial debonding and subsequent fibre pull-out.

subsequently pulled out of their sockets in the matrix. The work done when a single fibre undergoes interfacial debonding can be written as

$$\Delta U = 2\pi r x_0 G_{ic} \tag{9.12}$$

where x_0 is the embedded length of the fibre concerned on the side of the crack where debonding occurs ($x_0 \leq L$). To obtain the local work of debonding for the composite, G_{cd}, this is summed over all of the fibres intersected by the crack. If there are N fibres per m^2, then there will be $(N\,dx_0/L)$ per m^2 with an embedded length between x_0 and $(x_0 + dx_0)$. The total work done in debonding is therefore given by

$$G_{cd} = \int_0^L \frac{N\,dx_0}{L} 2\pi r x_0 G_{ic} \tag{9.13}$$

The value of N is related to the fibre volume fraction, f, and the fibre radius, r

$$N = \frac{f}{\pi r^2} \tag{9.14}$$

Substituting this expression into Eqn (9.13) and integrating leads to

$$G_{cd} = \frac{f 2\pi r G_{ic}}{\pi r^2 L} \frac{L^2}{2}$$

which can be rearranged to

$$G_{cd} = f s G_{ic} \tag{9.15}$$

Contributions from this mechanism are relatively small. For example, if $s = 50, f = 0.5$ and $G_{ic} = 10\,\mathrm{J\,m^{-2}}$, then $G_{cd} = 0.25\,\mathrm{kJ\,m^{-2}}$. If either the fibre aspect ratio or the interfacial fracture energy are much greater than these values ($s > s_*$ or $G_{ic} > G_{fc}$), then it is probable that the fibres will fracture in the crack plane and little or no debonding will occur. Slightly larger values are achievable in bending, when debonding can propagate for long distances from the fracture plane without causing the reinforcement to break. Recent work on ceramic laminates (Phillipps *et al.* 1993) has been based on this effect.

9.2.4 Frictional sliding and fibre pull-out

Potentially the most significant source of fracture work for most fibre composites is interfacial frictional sliding. Depending on the interfacial roughness, contact pressure and sliding distance, this process can absorb large quantities of energy. The case of most interest is pull-out of fibres from their sockets in the matrix. Fracture surfaces illustrating various extents of fibre pull-out are shown in Figs. 8.7–8.10. Pull-out aspect ratios can range up to several tens or even hundreds.

The work done can be calculated using a similar approach to that in §9.2.3. Consider a fibre with a remaining embedded length of x being pulled out an increment of distance $\mathrm{d}x$. The associated work is given by the product of the force acting on the fibre and the distance it moves

$$\mathrm{d}U = (2\pi r x \tau_{\mathrm{i}^*})\,\mathrm{d}x \qquad (9.16)$$

where τ_{i^*} is the interfacial shear stress, taken here as constant along the length of the fibre. The work done in pulling this fibre out completely is therefore given by

$$\Delta U = \int_0^{x_0} 2\pi r x \tau_{\mathrm{i}^*}\,\mathrm{d}x = \pi r x_0^2 \tau_{\mathrm{i}^*} \qquad (9.17)$$

The next step is a similar integration (over all of the fibres) to that used to obtain Eqn (9.13), leading to an expression for the pull-out work of fracture, G_{cp}

$$G_{\mathrm{cp}} = \int_0^L \frac{N\,\mathrm{d}x_0}{L} \pi r x_0^2 \tau_{\mathrm{i}^*} \qquad (9.18)$$

which, using Eqn (9.14) again for N, simplifies to

$$G_{\mathrm{cp}} = \frac{fs^2 r \tau_{\mathrm{i}^*}}{3} \qquad (9.19)$$

This contribution to the overall fracture energy can be large. For example, taking $f = 0.5$, $s = 50$, $r = 10\,\mu\mathrm{m}$ and $\tau_{\mathrm{i}^*} = 20\,\mathrm{MPa}$ gives a value of about $80\,\mathrm{kJ\,m^{-2}}$. Since σ_{f^*} would typically be about $3\,\mathrm{GPa}$, the critical aspect ratio, s_* ($= \sigma_{\mathrm{f}^*}/2\tau_{\mathrm{i}^*}$), for this value of τ_{i^*}, would be about 75. Since this is greater than the actual aspect ratio, pull-out is expected to occur (rather than fibre fracture), so the calculation should be valid. The pull-out energy is greater when the fibres have a larger diameter, assuming that the fibre aspect ratio is the same.

The potential for contribution to the toughness from fibre pull-out is substantial. However, the mechanism can apparently only operate with relatively short fibres ($s < s_*$). Continuous fibres are expected to break in the crack plane, since there will always be embedded lengths on either side of the crack plane which are long enough for the stress in the fibre to build up sufficiently to break it. (Interfacial debonding can still occur, perhaps over an appreciable distance, but it is clear from §9.2.3 that this is unlikely to make a major contribution to the fracture energy.) Using this argument, it appears that pull-out should not occur in long-fibre composites. In fact, it is often highly significant in such materials. The reason for this is related to the variability in strength exhibited by most fibres (see §2.2.4).

The effect of the variability in fibre strength, characterised by the Weibull modulus, m, is depicted schematically in Fig. 9.6. Figure 9.6(a)

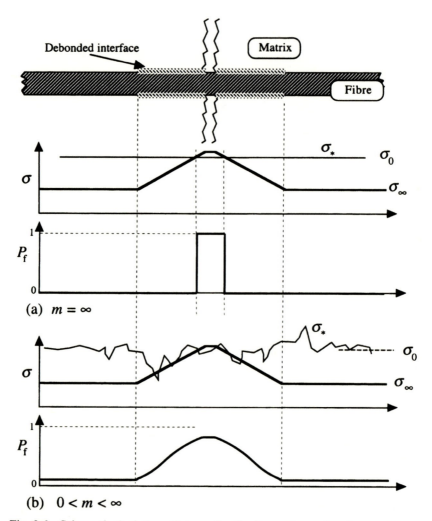

Fig. 9.6   Schematic depiction of stress distribution and associated probability of failure P_f along the length of a long fibre bridging a matrix crack. (a) A fixed fibre strength ($m = \infty$), and (b) a fibre strength which varies along its length (finite m).

shows how, in the case of a deterministic (single-valued) fibre strength ($m = \infty$), the probability, P_f, of the fibre breaking in the crack plane will eventually become 100%, while remaining zero elsewhere. For a finite value of m, however, (Fig. 9.6(b)) not only are high values of P_f spread over an appreciable distance on either side of the crack plane, but it is now possible for the fibre to break almost anywhere. Prediction of the fracture energy becomes more difficult under these circumstances, since account should be taken of these probabilities in calculating the contributions from different pull-out lengths. In fact, strictly, the effect of this distribution of fracture probabilities should also be taken into account for the short-fibre case. The details of such treatments need not concern us here, but it may be noted that a low Weibull modulus always tends to raise the pull-out work for long fibres, but is not necessarily beneficial for short ones. Detailed mapping of the various contributions to the overall toughness, for different Weibull moduli, has been undertaken by Wells and Beaumont (1988).

9.2.5 Effects of microstructure

There is considerable scope for controlling the fracture energy of composites by altering various features of the fibre length, orientation and interfacial characteristics. Examples of the effect of orientation are shown in Fig. 9.7. This shows measured fracture energies, obtained using the Charpy impact test. The fracture energy of the unidirectional lamina (Fig. 9.7(a)) falls off sharply as the angle between the crack plane and the fibre axis is reduced (loading angle increases). This is largely because fibre pull-out becomes inhibited and fracture occurs parallel to the fibre axis. The crossply laminate and woven cloth reinforced material exhibit more isotropic behaviour. It is difficult to predict the dependence on loading angle in these cases, since complex interactions occur between the different plies and fibre tows. However, fracture always involves a considerable degree of interfacial debonding, fibre fracture and fibre pull-out, so that the toughness is always relatively high. This is an important attribute of long-fibre laminates. Fig. 9.7(b) shows that the toughness of some metals can also be enhanced in this way. (The fracture energy of the matrix in this case was about $90 \, \text{kJ m}^{-2}$.) This figure also illustrates that good toughness can be obtained when the crack propagation direction lies in the plane of the laminate, provided the crack front is normal to this plane.

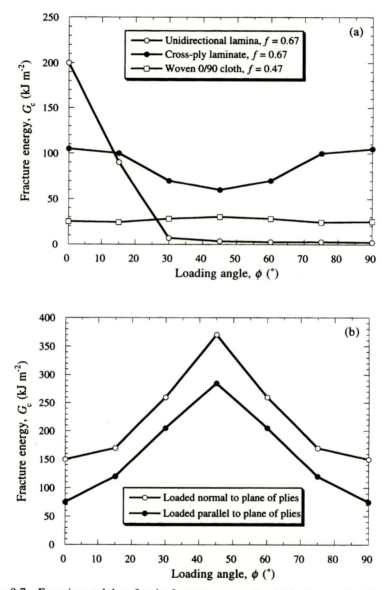

Fig. 9.7 Experimental data for the fracture energy G_c obtained using the Charpy impact test, as a function of the loading angle ϕ between the axis of the test bar and the reference fibre direction. (a) Epoxy resin/glass fibre, loaded normal to the laminate plane (Harris 1980), and (b) crossply laminates of Zn–8Al–1Cu/50% carbon fibre, loaded normal and parallel to the laminate plane (Vescera *et al.* 1991).

Control over interfacial properties is often of critical importance. This is illustrated by the data shown in Fig. 9.8. This refers to a ceramic material composed of sheets of SiC, separated by thin graphitic interlayers. When loaded in bending so that a crack propagates through successive sheets, deflection and extensive interfacial debonding occurs at each interlayer, provided the interfacial toughness is sufficiently low (see §9.1.2). Depending on the loading geometry and specimen dimension, the total work done, W, (= area under load–displacement plot) may exceed the actual energy absorbed in the specimen (the excess being dissipated through the loading system). Modelling of the interfacial cracking allows the true G_c to be established. It can be seen from Fig. 9.8 that raising the interfacial toughness (by changing the processing conditions), increased G_c, even though the total work done was little affected. Further increases in G_c could not, however, be achieved in this way, since a tougher interface (higher G_{ic}) would inhibit crack deflection. In this case, although toughening comes only from interfacial debonding, the fracture surface areas involved are such that the fracture energy values are relatively high for a ceramic material.

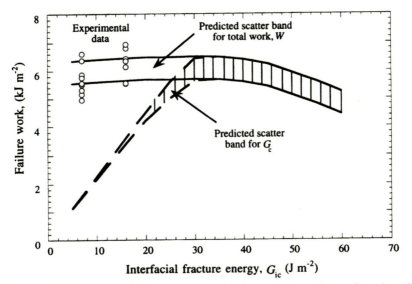

Fig. 9.8 Comparison between predicted and observed dependence of total work done W on interfacial fracture energy for SiC laminates containing graphitic interlayers. Also shown is the predicted dependence of the true fracture energy G_c which in this case is difficult to measure directly. (Phillipps *et al.* 1993).

For particulate-reinforced composites, the energy absorption from debonding and pull-out tends to be small. In particulate MMCs, a major concern is the retention of as much as possible of the inherent toughness of the matrix. Data are shown in Fig. 9.9 for the fracture energy of an age-hardening aluminium alloy containing SiC particles, as a function of the tensile strength. As ageing progresses, the tensile strength rises and then falls (over-ageing). The fall in fracture energy during the initial rise in strength is expected on the grounds of reduced dislocation mobility. The failure of the toughness to rise again as over-ageing occurs is more difficult to explain, but may be owing to precipitates forming preferentially at the interface, making it a preferred site for cavitation.

Finally, evaluation of the energy absorbed can be particularly complex for laminates, for which contributions can come from interlaminar debonding as well as intraply processes. A detailed discussion of the various interactions which can occur, and the significance of microstructural characteristics, is given by Hull and Shi (1993).

9.3 Sub-critical crack growth

When the rate of energy release during crack propagation (driving force) is lower than the critical value, spontaneous fast fracture does not occur.

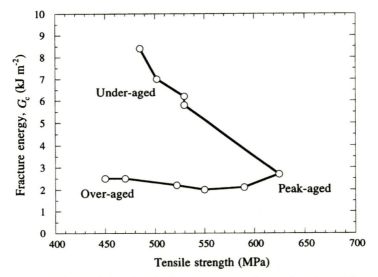

Fig. 9.9   Experimental data for the fracture energy G_c as a function of the tensile strength during progressive ageing of a 7000-series Al alloy reinforced with 15% of SiC particles (Lewandowski *et al.* 1989).

Under some circumstances, however, an existing crack may advance slowly under this driving force. Since crack growth leads to an increased driving force (for the same applied load), this process is likely to lead to an accelerating rate of damage, culminating in conditions for fast fracture being satisfied. There are two common situations in which such sub-critical crack growth tends to occur. Firstly, if the applied load is fluctuating in some way, local conditions at the crack tip may be such that a small advance occurs during each cycle. Secondly, the penetration of a corrosive fluid to the crack tip region may lower the local toughness and allow crack advance at a rate determined by the fluid penetration kinetics or chemical interaction effects. In both cases, as indeed with fast fracture, the presence or absence of an initial flaw, which allows the process to initiate, is of central importance.

9.3.1 Fatigue

For metals, fatigue failure is an important topic which has been the subject of detailed investigation. Analysis is commonly carried out in terms of the difference in stress intensity factor between the maximum and minimum applied load (ΔK). This is because, while the maximum value, K_{max}, dictates when fast fracture occurs, the cyclic dissipation of energy is dependent on ΔK. It is, however, common to quote the *stress ratio, R* ($= K_{min}/K_{max}$), which enables the magnitudes of the K values to be established for a given ΔK. The resistance of a material to crack extension is given in terms of the crack growth rate per loading cycle ($\mathrm{d}c/\mathrm{d}N$). At intermediate ΔK, the crack growth rate usually conforms to the Paris–Erdogan relation (Paris and Erdogan, 1963)

$$\frac{\mathrm{d}c}{\mathrm{d}N} = \beta(\Delta K)^n \tag{9.20}$$

where β is a constant. Hence, a plot of crack growth rate (m/cycle) against ΔK, with log. scales, gives a straight line in the Paris regime, with a gradient equal to n. At low stress intensities, there is a threshold, ΔK_{th}, below which no crack growth occurs. The crack growth rate usually accelerates as the level for fast fracture, K_c, is approached.

An alternative way of presenting fatigue data is in the form of S/N_f curves, showing the number of cycles to failure (N_f) as a function of the stress amplitude (S). Many materials show rapid crack growth (low N_f) when the stress amplitude is high, a central portion of decreasing S with rising N_f, corresponding to the Paris regime, and a *fatigue limit*, which is a stress amplitude below which failure does not occur even after large numbers of cycles. This corresponds to a stress intensity below ΔK_{th}.

Some of the features of fatigue can be illustrated by examining how the presence of particulate reinforcement affects the behaviour of a metal. This is shown schematically in Fig. 9.10. Below ΔK_{th}, cracks are unable to grow at all. For Al/SiC$_p$ MMCs, ΔK_{th} is typically around 2–4 MPa $\sqrt{m}$, which is approximately twice that for unreinforced Al alloys ($\sim$ 1–2 MPa $\sqrt{m}$). A number of explanations for this have been proposed (Christman and Suresh 1988), including crack deflection at interfaces and a reduction in slip-band formation due to the particles. This beneficial effect of the reinforcement in inhibiting the onset of fatigue cracks is a useful feature of such MMCs. However, for reasons outlined earlier (§9.2.1 and Table 9.1), MMCs have a lower fracture toughness than unreinforced metals, mainly as a result of the constraint imposed on matrix plasticity. The Paris regime is usually short and the exponent n is often around 5–6, which is higher than those typical of unreinforced systems ($\sim$ 4). Fast fracture is initiated at lower ΔK values than for the unreinforced metal. In practical terms, this means that MMCs can offer improvements in fatigue performance compared with metals, providing that applied stress levels are low and/or flaw sizes are kept small. Control over processing of MMCs so as to eliminate inhomogeneities is important to ensure this.

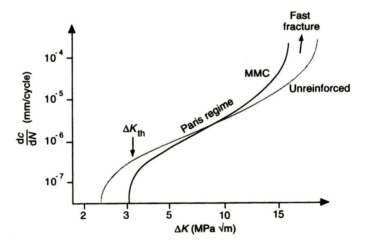

Fig. 9.10 Schematic depiction of fatigue crack growth rate as a function of applied stress intensity factor ΔK, for a typical particle-reinforced MMC and the corresponding unreinforced alloy, illustrating the effect of reinforcement on the fatigue response.

While particulate reinforcement produces relatively minor modifications to the behaviour of the matrix, the presence of long fibres has a more pronounced effect. This is particularly true for polymer composites, in which it is usually the type and orientation distribution of the fibres which is of most significance. The propagation of cracks through the matrix and along the interfaces usually dictates how fatigue progresses, but this is strongly influenced by how the fibres affect the stress distribution. The failure strain of the matrix is also important. A further difference from the particulate case lies in the distribution of damage. It is common in long-fibre composites for matrix and interfacial microcracks to form at many locations throughout the specimen. Fibre bridging across matrix cracks often occurs, reducing the stress intensity at the crack tip. In contrast to this, fatigue crack growth in monolithic and particle-reinforced materials usually involves a single dominant crack with a well-defined length.

The S/N_f curves presented in Fig. 9.11 illustrate some features of the fatigue performance of long-fibre reinforced polymers, when loaded along the fibre axis. Composites reinforced with stiff fibres, such as

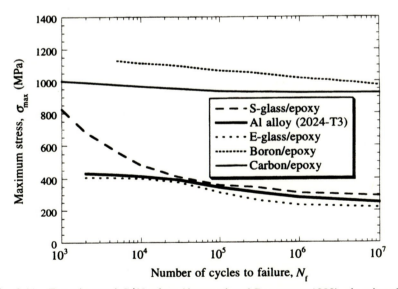

Fig. 9.11 Experimental S/N_f plots (Agarwal and Broutman 1980), showing the number of cycles to failure during axial fatigue loading of unidirectional long-fibre polymer composites, as a function of the peak applied load. The stress ratio R was 0.1 in all cases. Also shown is a plot for a typical unreinforced aluminium alloy.

boron and carbon, show excellent fatigue resistance, being able to withstand alternating loads of around 1 GPa for very large numbers of cycles.
The fatigue performance of these materials is markedly superior to that
of a typical aluminium alloy. With glass fibres, on the other hand, the
lower stiffness of the fibre leads to reduced stress transfer, exposing the
matrix to larger stresses and strains. This causes progressive damage at
considerably lower applied loads than for the stiffer fibres.

While the axial fatigue resistance of long-fibre composites tends to be
very good, particularly with high stiffness fibres, performance is usually
inferior for laminates or under off-axis loading. This is illustrated by the
plots in Fig. 9.12, which are for glass-reinforced polymer. The crossply
and woven cloth laminates fail at appreciably lower loads than the unidirectional material and show little evidence of a fatigue limit stress value
being identifiable. Damage to the transversely oriented regions starts at
low applied loads (see §8.1.2), transferring extra load and eventually
causing cracks to propagate into the axial regions. Nevertheless, the
fatigue resistance of such materials compares quite well with that of
many metals. Finally, the chopped strand mat (CSM) and dough

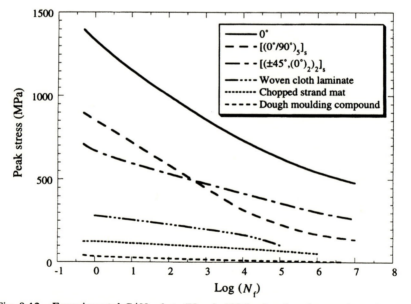

Fig. 9.12 Experimental S/N_f plots (Harris 1994), showing the number of cycles
to failure during fatigue loading of glass fibre/polyester composites with various
fibre distributions, as a function of the peak applied load. The stress ratio R
$(= \sigma_{max}/\sigma_{min})$ was 0.1 in all cases.

moulding compound (DMC) show relatively poor fatigue resistance. In these materials, fibres are misaligned and have relatively low aspect ratios, particularly for the DMC.

Broadly similar behaviour is exhibited by MMC laminates. Data in the form of S/N_f plots are shown in Fig. 9.13(a). The best performance is shown by the unidirectional material, having the fibres parallel to the applied load. The performance of the other laminates can be rationalised by calculating the range of stress to which the $0°$ ply (parallel to the applied load) is subjected during loading. When this is plotted against the number of cycles to failure (Fig. 9.13(b)), then the data for the different laminates fall on a common curve. This highlights the point that the laminate does not fail until the fibres in the $0°$ ply become fractured. The fatigue properties can, however, become badly degraded if matrix cracks are not deflected at the fibre/matrix interface (Johnson 1993).

A further point worthy of note with respect to the fatigue of composites is that the behaviour is often sensitive to the absolute values of the stresses being applied, rather than just the ΔK range. In particular, the introduction of compressive stresses usually reduces the resistance to fatigue. This is largely due to the axially aligned fibres having poor resistance to buckling (see §2.2.2). This results in damage to the fibres and the surrounding matrix and also allows larger stresses to bear on neighbouring, transversely oriented regions, accelerating their degradation. Aramid (e.g. Kevlar^{TM}) fibres are particularly prone to this effect, since they have poor resistance to compression. This is illustrated by the data in Fig. 9.14, which show that the fatigue resistance of Kevlar^{TM} composites becomes poor for negative R values. (It should, however, be noted that the effect is exaggerated by plotting the peak stress rather than the stress difference, which is larger for the lower R values.) Large-diameter monofilaments, on the other hand, tend to be resistant to buckling, leading to improved performance at lower R values.

A final point concerns the frequency of cycling. For metals, this usually has little effect, but in polymer composites a higher frequency often hastens failure. This is partly because of the viscoelastic response of polymers. Matrix damage is more likely if the strain is imposed rapidly, allowing no time for creep and stress relaxation. A second effect arises from the low thermal conductivity of polymers, particularly if the fibres are also poor conductors (glass, Kevlar^{TM}, etc.). Such composites tend to increase in temperature during fatigue loading, as a result of difficulties in dissipating the heat generated locally by damage and viscoelastic deformation. This is accentuated at high cycling frequencies. Since the strength

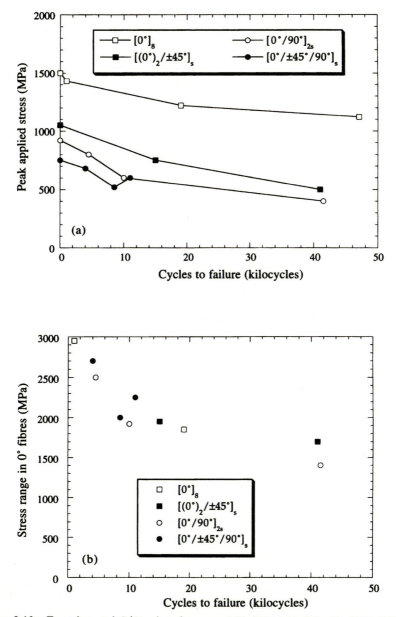

Fig. 9.13   Experimental S/N_f plots for several Ti–15V–3Al–3Cr–3Sn/35% SCS-6 SiC monofilament laminate composites (Johnson 1993). (a) Applied stress against number of cycles to failure, and (b) the same data plotted as the stress range experienced by the fibres in the 0° plies against number of cycles to failure.

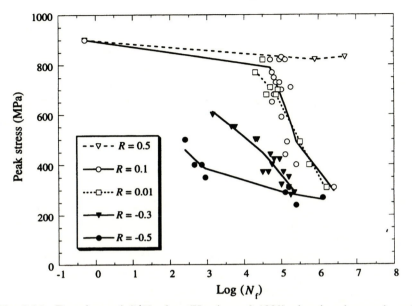

Fig. 9.14 Experimental S/N_f plots (Harris *et al.* 1990), showing the number of cycles to failure during axial fatigue loading of unidirectional Kevlar™ fibre composites, with various values of the stress ratio R $(= \sigma_{max}/\sigma_{min})$, as a function of the peak applied load.

of the matrix falls with increasing temperature, this tends to accelerate failure.

9.3.2 Stress corrosion cracking

Stress corrosion cracking is the term given to sub-critical crack growth which occurs as a result of the effect of a corrosive fluid reaching the advancing crack tip. The micromechanisms responsible for this lowering of the local toughness of the material vary widely between different materials and environments. For example, in Al-based systems, the presence of moist or salt-laden air causes an acceleration of fatigue crack growth of between 5–10 times. A similar effect is observed for particulate MMCs (Crowe and Hasson 1982). For long-fibre reinforced aluminium, the fibres give rise to good retention of fatigue resistance under axial loading, whereas in the transverse direction the fatigue resistance is affected by the corrosive environment to a degree directly related to the proportion of matrix occupied by the composite (Hasson and Crowe 1987).

In the case of aluminium, stress corrosion cracking is usually due to the evolution of (atomic) hydrogen at the crack tip, which is promoted by the presence of various liquids. This tends to cause embrittlement by impeding dislocation motion. In MMCs, a further factor is introduced in the form of strong traps for hydrogen at the matrix/reinforcement interface, promoting cracking. Using straining electrode tests, Bernstein and Dollar (1990) showed that, while the strength levels of MMCs were unaffected by testing in a hydrogen environment, the ductility was sharply reduced, irrespective of matrix ageing. This was attributed to strong trapping of hydrogen at the interface, where voids (not seen in air-tested specimens) were observed.

For polymer composites, the behaviour of the matrix is often quite sensitive to the presence of fluids. For example, the properties of many polymers are modified by absorption of moisture. In many cases, this penetration occurs by diffusion through the matrix and is not confined to access along the crack. Various features of such absorption have been identified (Shen and Springer 1976). Commonly, water absorption raises the toughness and strain to failure of the matrix. Thus, the fatigue resistance of glass-reinforced epoxy can be raised by prior boiling in water (Harris 1994). However, water uptake also tends to promote interfacial debonding, impairing stiffness and strength, particularly under shear and transverse loading. Some fluids sharply degrade the strength of the fibres. An example of this can be seen in Fig. 8.10, which shows the fracture surface of a polyester/60% glass composite tested in the presence of dilute hydrochloric acid. This attacks the fibre surface at the crack tip, reducing its strength considerably. Not only does this impair the strength of the composite, but the toughness and fatigue resistance are sharply reduced, since the fibres all break in the crack plane and pull-out work (§9.2.4) is virtually zero. The danger of such effects must always be borne in mind when glass-reinforced polymer composites are to be used under load in chemically aggressive environments, particularly mineral acids (Price and Hull 1987).

References and further reading

Agarwal, B. D. and Broutman, L. J. (1980) *Analysis and Performance of Fiber Composites.* Wiley: New York

Akisanya, A. R. and Fleck, N. A. (1992) Brittle fracture of adhesive joints, *Int. J. Fracture,* **58** 93–114

Ashby, M. F. and Jones, D. R. H. (1980) *Engineering Materials.* Pergamon: Oxford

Bernstein, I. M. and Dollar, M. (1990) The effect of trapping on hydrogen-induced plasticity and fracture in structural alloys, in *Hydrogen Effects on Material Behaviour*. N. R. Moody and A. W. Thompson (eds.) TMS pp. 703–15

Christman, T. and Suresh, S. (1988) Effects of SiC reinforcement and ageing treatment on fatigue crack growth in an Al–SiC composite, *Mat. Sci. & Eng.*, **102A** 211–16

Cook, J. and Gordon, J. E. (1964) A mechanism for the control of crack propagation in all-brittle systems, *Proc. Roy. Soc.*, **282A** 508–20

Crowe, C. R. and Hasson, D. F. (1982) Corrosion fatigue of SiC/Al MMC in salt-laden moist air, in *Proc. ICMSA 6*. R. C. Gifkins (ed.) Pergamon pp. 859–65

Evans, A. G. and Hutchinson, J. W. (1989) Effects of non-planarity on the mixed mode fracture resistance of bimaterial interfaces, *Acta Metall. Mater.*, **37** 909–16

Evans, A. G., Rühle, M., Dalgleish, B. J. and Charalambides, P. G. (1990) The fracture energy of bimaterial interfaces, *Mat. Sci. & Eng.*, **A126** 53–64

Ewalds, H. L. and Wanhill, R. J. H. (1984) *Fracture Mechanics*. Edward Arnold: London

Gordon, J. E. (1978) *Structures*. Penguin: New York

Griffith, A. A. (1920) The phenomena of rupture and flow in solids, *Phil. Trans. Roy., Soc.* **A221** 163–97

Harris, B. (1980) *Engineering Composite Materials*. Inst. of Metals: London

Harris, B. (1994) Fatigue – glass fibre reinforced plastics, in *Handbook of Polymer Fibre Composites*, F. R. Jones (ed.) Longman pp. 309–16

Harris, B., Reiter, H., Adam, T., Dickson, R. F. and Fernando, G. (1990) Fatigue behaviour of carbon fibre reinforced plastics, *Composites*, **21** 232–42

Hasson, D. F. and Crowe, C. R. (1987) Flexural fatigue behavior of aramid reinforced Al 7075 laminate and Al 7075 Sheet in air and in salt-laden humid air, in *Proc. ICCM 6*, F. L. Matthews *et al.* (eds.) Elsevier pp. 138–45

He, M. Y. and Hutchinson, J. W. (1989) Crack deflection at an interface between dissimilar and elastic materials, *Int. J. Solids Structures*, **25** 1053–67

Hull, D. and Shi, Y. B. (1993) Damage mechanism characterisation in composite damage tolerance investigations, *Composite Structures*, **23** 99–120

Inglis, C. (1913) Stress in a plate due to the presence of cracks and sharp corners, *Trans. Inst. Naval Architects*, **55** 219–30

Irwin, G. R. (1948) Fracture dynamics, in *Fracture of Metals*. ASM pp. 152–69

Johnson, W. S. (1993) Damage development in titanium metal matrix composites subjected to cyclic loading, *Composites*, **24** 187–96

Kendall, K. (1975) Transition between cohesive and interfacial failure in a laminate, *Proc. Roy. Soc.*, **344A** 287–302

Lewandowski, J. J., Liu, C. and Hunt, W. H. (1989) Effect of matrix microstructure and particle distribution on fracture of an Al MMC, *Mat. Sci. & Eng.*, **A107** 241–55

Paris, P. and Erdogan, F. (1963) A critical analysis of crack propagation laws, *J. Basic Engg*, **85** 528–34

Phillipps, A. J., Clegg, W. J. and Clyne, T. W. (1993) Fracture of ceramic laminates in bending. Part II – Comparison of model predictions with experimental data, *Acta Metall. Mater.*, **41** 819–27

Price, J. N. and Hull, D. (1987) Effect of matrix on crack propagation during stress corrosion of glass reinforced composites, *Comp. Sci. Tech.*, **28** 193–210

Rice, J. R. (1988) Elastic fracture mechanics concepts for interfacial cracks, *J. Appl. Mech.*, **55** 98–103

Shen, C. H. and Springer, G. S. (1976) Moisture absorption and desorption of composite materials, *J. Comp. Mat.*, **10** 2–20

Suo, Z. and Hutchinson, J. W. (1990) Interface crack between two elastic layers, *Int. J. Fracture*, **43** 1–18

Vescera, F., Keustermans, J. P., Dellis, M. A., Lips, B. and Delannay, F. (1991) Processing and properties of Zn–Al matrix composites reinforced by bidirectional carbon tissues, in *Metal Matrix Composites – Processing, Microstructure and Properties*. N. Hansen *et al.* (eds.) Risø Nat. Lab.: Roskilde, Denmark pp, 719–24

Wang, J. S. and Suo, Z. (1990) Experimental determination of interfacial fracture toughness curves using Brazil nut sandwiches. *Acta Metall. Mater.*, **38** 1279–80

Wells, J. K. and Beaumont, P. W. R. (1988) The toughness of a composite containing short brittle fibres, *J. Mater. Sci.*, **23** 1274–82

10

Thermal behaviour of composites

The behaviour of composite materials is often sensitive to changes in temperature. This arises for two main reasons. Firstly, the response of the matrix to an applied load is temperature-dependent and, secondly, changes in temperature can cause internal stresses to be set up as a result of differential thermal contraction and expansion of the two constituents. These stresses affect the thermal expansivity (expansion coefficient) of the composite. Furthermore, significant stresses are normally present in the material at ambient temperatures, since it has in most cases been cooled at the end of the fabrication process. Changes in internal stress state on altering the temperature can be substantial and may strongly influence the response of the material to an applied load. Creep behaviour is affected by this, particularly under thermal cycling conditions. Finally, the thermal conductivity of composite materials is of interest, since many applications and processing procedures involve heat flow of some type. This property can be predicted from the conductivities of the constituents, although the situation may be complicated by poor thermal contact across the interfaces.

10.1 Thermal expansion and thermal stresses

10.1.1 Thermal stresses and strains

Data for the thermal expansion coefficients (α) of matrices and reinforcements, as a function of temperature, are shown in Fig. 10.1. Polymers and metals generally expand more than ceramics. It can be seen that the differences in expansivity between fibre and matrix are large in many cases. Since fabrication almost inevitably involves consolidation at relatively high temperatures (see Chapter 11), composites usually

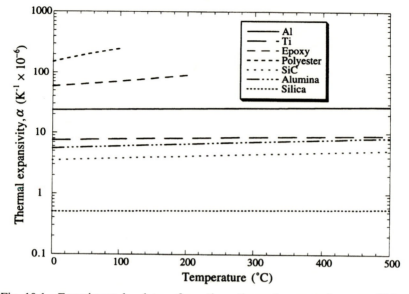

Fig. 10.1 Experimental data for the thermal expansion coefficients (expansivities) of various reinforcements and matrices.

contain significant differential thermal contraction stresses at ambient temperatures. In addition, for resin-based composites, the matrix undergoes shrinkage during curing, which represents a further contribution to differential contraction. Assuming the material to be effectively stress-free at some (high) temperature, T_{esf}, the stress state at a lower temperature can be envisaged as arising from the fitting of an oversized inclusion into an undersized hole in the matrix. The ***misfit strain*** is then $\Delta\alpha\Delta T$, where $\Delta T = T_{esf} - T_0$ (ambient temperature) and $\Delta\alpha = \alpha_m - \alpha_f$.

For certain cases, the stress state which results from the misfit strain can be calculated analytically. The simplest situation is when the reinforcement is an isolated sphere in an infinite matrix. The stresses in the matrix are the same as those produced by a spherical bubble of radius a under a pressure P ($=$ hydrostatic stress, σ_H of $-P$). Lamé (1852) first obtained the solution for the radial and hoop (tangential) stresses, σ_r and σ_θ, in the matrix for this case

$$\sigma_r = -\frac{Pa^3}{r^3}$$

$$\sigma_\theta = \frac{Pa^3}{2r^3}$$

$$(10.1)$$

The pressure in a spherical particle as a result of a misfit strain of $\Delta\alpha\Delta T$ can also be expressed analytically (Lee *et al.* 1980).

$$P = \frac{4G_m \dfrac{(1 + \nu_m)}{3(1 - \nu_m)} \dfrac{K_p}{K_m} \Delta\alpha\Delta T}{\dfrac{(1 + \nu_m)}{3(1 - \nu_m)} \left(\dfrac{K_p}{K_m} - 1\right) + 1} \qquad (10.2)$$

where G, ν and K are shear modulus, Poisson's ratio and bulk modulus respectively, and the subscripts p and m refer to particle and matrix, respectively. The result of applying these equations to the case of a SiC sphere in a titanium matrix after cooling through 500 deg K is shown in Fig. 10.2(a). This shows that a large compressive stress is developed in the radial direction, but the hoop stress in the matrix is tensile. There is, therefore, a large deviatoric (shape-changing) component to the stress state in the matrix near the sphere. In this case, the stresses are likely to cause yielding of the matrix. Equations are also available (Lee *et al.*

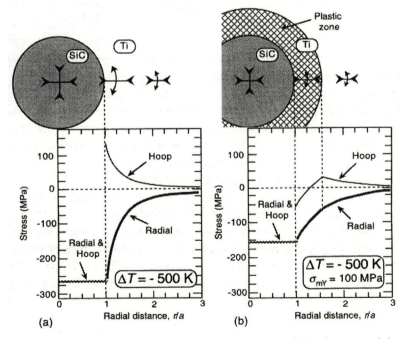

Fig. 10.2 (a) Elastic stress field in an infinite matrix of titanium containing a spherical particle of SiC, after cooling through 500 K. (b) New stress state after local plastic flow (Lee *et al.* 1980), assuming a matrix yield stress of 100 MPa, with no work hardening.

1980) for prediction of the extent of the plastic zone and the associated changes in stress distribution. These are shown in Fig. 10.2(b). It can be seen that the plastic flow has caused substantial changes in the stress distribution, including a reduction in the hydrostatic pressure to which the particle is subjected. While the presence of neighbouring particles does affect the situation in a real composite, the fact that matrix plasticity often changes the stress state predicted for the elastic case remains true and should always be borne in mind.

The elastic stress field for an infinitely long single fibre surrounded by a cylindrical matrix (of finite radial extent) can also be obtained analytically, although the equations involved are more complex. A typical result was shown in Fig. 7.3, for a glass/polyester composite after cooling through 100 K. This stress field shows similarities to the case of the sphere, although, as might have been expected, the axial stresses become prominent for the fibre case. In a short-fibre composite, the stress field is more complex, particularly when the effect of neighbouring fibres is considered. In this case, recourse to numerical methods, with considerable computational effort, is necessary to predict the stress field. However, the general nature of the stress distribution, and some idea of the magnitudes of the stresses, can be inferred from these simpler cases. It is also possible to use the Eshelby method (§6.2) to predict the volume-averaged matrix stresses resulting from differential thermal contraction.

Before going on to examine how thermally induced stresses can affect inelastic deformation processes, it is useful to consider how the elastic accommodation of the strain misfit controls the overall coefficient of thermal expansion of the composite.

10.1.2 *Thermal expansivities*

Analysis of the stresses due to a change in temperature allows the coefficient of thermal expansion (CTE) to be predicted. These stresses will have associated strains and the net effect of these on the length of the composite, in any given direction, can be calculated or estimated. This net length change arising from the internal stresses is simply added to the natural thermal expansion of the matrix to give the overall length change and hence the composite expansivity. This simple view of thermal expansion allows certain points to be identified immediately. For example, a porous material, regarded as a composite of voids in a matrix, does not develop any internal stresses on heating because the 'inclusions' have zero stiffness. Hence the presence of pores (of whatever shape, size and volume fraction) does not affect the CTE, although they give rise to sharp reductions in stiffness.

The simplest case to treat is the axial expansion of a long-fibre composite. It can be seen from Fig. 7.3 that the axial stresses are uniform in both constituents, which simplifies the problem. The slab model provides a good basis for this calculation, as it does for the Young's modulus (§4.1). There are two stages in the calculation, illustrated in Fig. 10.3. Firstly, both constituents are allowed to expand freely. Then, they are constrained to have the same length in the axial direction, by the imposition of appropriate stresses. The final net expansion can then be treated as consisting of two terms; a natural expansion (that of the matrix is

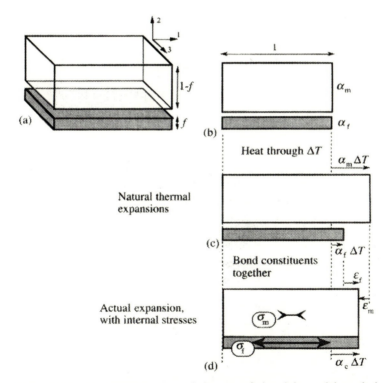

Fig. 10.3 Schematic representation of the use of the slab model to derive an expression for the axial expansivity of a long-fibre composite. (a) Slab dimensions, (b) initial length in the fibre direction, (c) unconstrained thermal expansions of the two constituents, and (d) final dimensions and internal stresses, when interfacial bonding is maintained.

usually considered) and an additional contraction or expansion caused by the internal stress. Thus

$$\alpha_c \Delta T = \alpha_m \Delta T + \epsilon_m \tag{10.3}$$

where α_c is the expansivity of the composite and ϵ_m is the elastic strain in the matrix associated with the internal stress (see Fig. 10.3). The value of ϵ_m is found by using two simple relationships, based on a strain balance and a force balance. The *strain balance* is obtained by expressing the sum of the two internal stress-generated strains as the difference between the natural thermal strains

$$(\alpha_m - \alpha_f)\Delta T = \epsilon_f - \epsilon_m \tag{10.4}$$

The *force balance* is simply a requirement that, since there is no applied stress, the two internal stresses must counterbalance each other when they are converted to forces (multiplied by the relative sectional areas)

$$\begin{aligned} (1-f)\sigma_m + f\sigma_f = 0 \\ (1-f)E_m\epsilon_m + f E_f\epsilon_f = 0 \end{aligned} \tag{10.5}$$

Combining Eqns (10.4) and (10.5) to derive an expression for ϵ_m

$$(1-f)E_m\epsilon_m + f E_f[\epsilon_m + (\alpha_m - \alpha_f)\Delta T] = 0$$
$$\therefore \epsilon_m = \frac{-f E_f(\alpha_m - \alpha_f)\Delta T}{(1-f)E_m + f E_f} \tag{10.6}$$

Substitution of this into Eqn (10.3) allows an expression to be obtained for the (axial) thermal expansivity of the composite

$$\begin{aligned} \alpha_c &= \alpha_m - \frac{f E_f(\alpha_m - \alpha_f)}{(1-f)E_m + f E_f} \\ &= \frac{\alpha_m(1-f)E_m + \alpha_f f E_f}{(1-f)E_m + f E_f} \end{aligned} \tag{10.7}$$

Since the axial force balance is reliable for the slab model (§4.1), this prediction should be quite accurate. (It is not entirely rigorous, because differential Poisson contraction strains are neglected.)

The expansivity in the transverse direction, and the values for short-fibre and particulate composites, are more difficult to establish, since the stresses and strains vary with position. Nevertheless, as with the transverse stiffness, some useful approximations can be made. A review of the models developed in this area is given by Bowles and Tompkins (1989). One of the most successful for the transverse expansivity of long-fibre composites is that due to Schapery (1968), who used an energy-based approach to obtain the following expression

$$\alpha_c^{tr} = \alpha_m(1 - f)(1 + \nu_m) + \alpha_f f(1 + \nu_f) - \alpha_c^{ax}\nu_{12c} \qquad (10.8)$$

in which the axial expansivity, α_c^{ax}, is that given by Eqn (10.7) and the Poisson ratio ν_{12c} is obtained from a simple rule of mixtures between those of the constituents (§4.4).

The predicted dependence of these two expansivities on fibre content, given by Eqns (10.7) and (10.8), is shown in Fig. 10.4 for a polymer matrix composite. Also shown are predictions obtained using the Eshelby method (§6.2). The core of this technique involves calculating the volume-averaged stresses in both constituents as a result of an imposed misfit strain, so that it is readily adapted for the prediction of expansivities. The equation obtained (Clyne and Withers 1993) is

$$\alpha_c = \alpha_m - f\{(C_m - C_f)[S - f(S - I)] - C_m\}^{-1}C_f(\alpha_f - \alpha_m) \qquad (10.9)$$

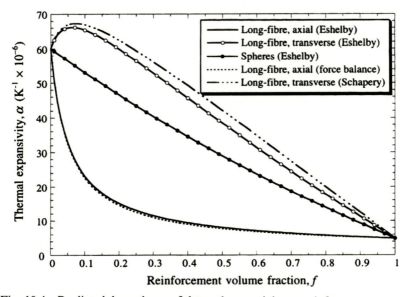

Fig. 10.4 Predicted dependence of thermal expansivity on reinforcement content for glass fibres and spheres in an epoxy matrix, according to the force balance (Eqn (10.7)), Schapery (Eqn (10.8)) and Eshelby (Eqn (10.9)) models. The curves were obtained using the property data in Tables 2.2 and 2.5.

where the stiffness, C, and expansivity, α, are here both tensors, as are the Eshelby tensor, S, (which depends on the aspect ratio of the fibre) and the identity tensor, I.

Several features of the curves in Fig. 10.4 are worthy of note. Firstly, the force balance (Eqn (10.7)) and Schapery (Eqn (10.8)) predictions agree well with the Eshelby model, confirming that they are quite reliable. Secondly, the case of spherical particles is quite well represented by a rule of mixtures (linear variation between the expansivities of the constituents). Finally, the transverse expansivity of a (long-fibre) composite tends to rise initially as the fibre content increases. This occurs because, on heating the composite, axial expansion of the matrix is strongly inhibited by the presence of the fibres and the resultant axial compression of the matrix generates a Poisson expansion in the transverse direction, which more than compensates for the reduction effected in the normal way by the presence of the fibres, at least for low fibre contents.

Plots such as those in Fig. 10.4 are of interest to engineers, since they allow the tailoring of expansivity via selection of constituents and reinforcement contents. However, it is important to note that these values are based on elastic behaviour. The associated internal stresses may become large, particularly if the temperature changes involved are substantial, and under these circumstances the matrix is likely to undergo plastic flow, or creep, which will alter the dimensional changes exhibited by the composite and make them difficult to predict.

10.1.3 Thermal cycling of unidirectional composites

Large internal stresses can be generated when the temperature of a composite is changed. This often occurs during service, since temperature fluctuations of at least several tens of °C are likely even with components which are not designed for high- (or low-) temperature use. The behaviour of composites during such thermal excursions is therefore of practical importance. Composites may respond to the associated internal stress changes in an inelastic manner. For example, dilatometry (length) measurements made on composites often exhibit significant hysteresis (i.e. the heating and cooling curves do not coincide).

A schematic representation of the effects of thermal cycling is presented in Fig. 10.5. This shows changes in axial matrix stress and composite strain over a relatively large temperature range, for a matrix prone to yielding (e.g. a metal or a thermoplastic). The matrix is initially (point A) taken as having a residual tensile stress equal to the yield stress

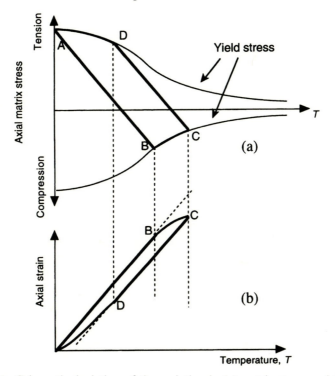

Fig. 10.5 Schematic depiction of the variation in (a) matrix stress and (b) axial strain of the specimen during thermal cycling of an aligned fibre composite.

(because of the cooling cycle). On heating, this stress falls, becomes compressive and eventually causes yielding in compression (point B). A period of progressive plastic flow then follows. On cooling (from point C), the matrix yields in tension at point D, before returning to A. The dilatometry traces (axial strain–temperature plots) are predicted to show hysteresis, but no net dimensional change.

Experimental data for long-fibre composites (Kural and Min 1984, Wolff *et al.* 1985) are broadly consistent with this view. A simple analysis of the dilatometer trace, proposed by Masutti *et al.* (1990), can be used to estimate the changing axial matrix stress in a long-fibre composite. Assuming no interfacial sliding, fibre yielding, etc., the axial strain of the composite must be equal to that of the fibres, which can be expressed as the sum of their natural thermal expansion and their elastic strain

$$\epsilon_{1c} = \epsilon_{1f} = \alpha_f \Delta T + \frac{\sigma_{1f} - \nu_f(\sigma_{2f} + \sigma_{3f})}{E_f} \tag{10.10}$$

Since the radial and hoop stresses in the fibre (σ_{2f} and σ_{3f}) are relatively small compared with the axial stress (see Fig. 7.3) and the Poisson ratio of ceramic fibres fairly low (~ 0.2), the contribution from the transverse fibre stresses can be neglected. The axial force balance ($f\sigma_{1f} + (1-f)\sigma_{1m} = 0$) can then be used to find the axial stress in the matrix

$$\sigma_{1m} = \frac{f}{(1-f)} E_f(\alpha_f \Delta T - \epsilon_{1c}) \tag{10.11}$$

Hence, σ_{1m} is simply proportional to the difference, $\Delta\epsilon$, between the natural thermal strain of the fibre and the measured strain of the composite. The initial thermal stress in the matrix can be deduced by taking the composite up to high temperature ($\sim 0.8T_m$), where the matrix stress will become very small, and running the fibre thermal expansion line back from this region. This is illustrated in Fig. 10.6, which shows an experimental dilatometer trace and the deduced matrix stress history. The latter is seen to be broadly of the form shown in Fig. 10.5. This simple procedure forms a convenient way of studying the initial matrix stress, as well as the high-temperature characteristics. For example, Masutti *et al.* (1990) showed that a quench in liquid nitrogen, followed by heating to room temperature, generated a compressive residual stress of about 25 MPa in an Al/SiC composite, compared with a tensile stress of 50 MPa after heating and cooling back to room temperature.

The behaviour of short-fibre composites during thermal cycling is rather similar to that described above, but there is more scope for various stress relaxation processes to operate. This is illustrated by the neutron diffraction data of Withers *et al.* (1987) presented in Fig. 10.7, which shows the changing axial strains (and hence stresses) in both fibre and matrix. The matrix stress history is similar to those shown in Figs. 10.5 and 10.6, although the observation of a significant drop in the final residual stress over an 8-hour period at room temperature is suggestive of continuing stress relaxation (creep) processes. The scope for such stress relaxation varies markedly with the type of matrix, the fibre length and the temperature range. When stress relaxation cannot occur easily, and the matrix is relatively brittle, then matrix microcracking is likely. This is more common with laminates than with unidirectional composites – see §10.1.4.

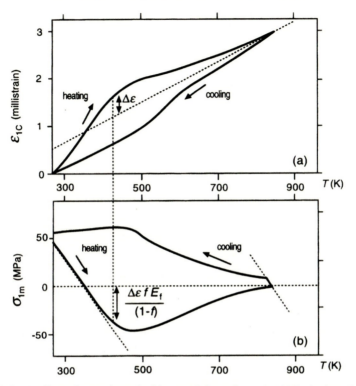

Fig. 10.6 (a) Experimental strain history (Masutti *et al.* 1990) during thermal cycling and (b) deduced variation of matrix axial stress for an Al–3Mg/30% SiC long-fibre composite. The matrix stress is taken, Eqn (10.11), as being proportional to the difference between the measured composite strain and the natural thermal expansion of the fibres, which can be obtained by extrapolation of the high-temperature data. The initial portions of the stress history curve, at the start of heating and of cooling, have gradients corresponding to elastic behaviour, indicated by the dotted lines.

10.1.4 Thermal cycling of laminates

Problems associated with thermal stresses can be severe with laminates. Thermal misfit strains now arise, not only between fibre and matrix, but between the individual plies of the laminate. For example, since a lamina usually has a much larger expansion coefficient in the transverse direction than in the axial direction, heating of a crossply laminate will lead to the transverse expansion of each ply being strongly inhibited by the presence of the other ply. This is useful in the sense that it will make the dimensional changes both less pronounced and more isotropic, but it does lead

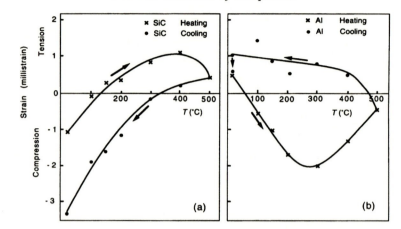

Fig. 10.7 Neutron diffraction measurements (Withers *et al.* 1987) of the lattice strain variation over a thermal cycle for Al containing 5% aligned SiC whiskers in (a) reinforcement and (b) matrix. The observed stress relaxation at the end of the cycle took place over a period of 8 hours.

to internal stresses in the laminate. These can cause the laminate to distort, so that it is no longer flat; an illustration of the type of distortion which can occur was shown in Fig. 5.14 (§5.4.3). However, even if such distortions do not occur, the stresses which arise on changing the temperature can cause microstructural damage and impairment of properties.

Thermal stresses within a laminate can be calculated using the numerical techniques outlined in Chapter 5. For the special case of a crossply laminate, a simple analytical approach can be used. The stresses and strains parallel and normal to the fibres in each ply can be found using the method outlined in §10.1.2. Figure 10.3 is again applicable, but the two constituents are now the two plies (one oriented axially and the other transversely), rather than the matrix and the fibres. The same equations result, but the meaning of the parameters changes slightly. For example, taking the transverse ply as the 'matrix' and the axial one as the 'fibres', a modified version of Eqn (10.6) now gives the elastic strain in the transverse ply on heating through a temperature interval ΔT

$$\epsilon_2 = \frac{-E_1(\alpha_2 - \alpha_1)\Delta T}{E_2 + E_1}$$

where E_1, E_2 are the axial and transverse Young's moduli (given by Eqns (4.3) and (4.7) respectively) and α_1, α_2 are the axial and transverse expansivities (given by Eqns (10.7) and (10.8) respectively). The volume fraction, f, of the two constituents does not appear, since it is equal to 0.5 for a crossply laminate in which the plies are of equal thickness and therefore cancels out in Eqn (10.6). This strain can be converted directly to a stress, since it arises from the two plies being forced to have the same length (and does not include the free thermal expansion). The stress to which the transverse ply is subjected can therefore be expressed as

$$\sigma_2 = \frac{-E_1 E_2 (\alpha_2 - \alpha_1)\Delta T}{E_2 + E_1} \tag{10.12}$$

The stress in the axial ply is equal in magnitude to this and opposite in sign.

Plots obtained using Eqn (10.12) are presented in Fig. 10.8. This shows the stresses in the transverse plies of two crossply laminates, as a function of the volume fraction of fibres, after cooling through a temperature interval of 100 deg K. Property data for the fibres and the matrix were taken from Tables 2.2 and 2.5. Several features are of interest in this

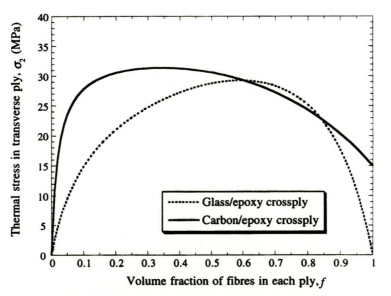

Fig. 10.8 Predicted dependence on fibre content of the thermal stress in the transverse direction within each ply of two crossply laminates, arising from a temperature decrease of 100 K. The plots were obtained from Eqn (10.12), using fibre data in Table 2.2 for E-glass and carbon (HM) and data for epoxy resin in Table 2.5.

figure. A peak is observed on increasing the fibre content. This is expected, since the stresses are due to the anisotropy in α; the difference between α_1 and α_2 has a peak at some intermediate value of f (see Fig. 10.4). For the carbon fibre composite, the stress does not fall to zero as f approaches 100%. This is because the carbon fibres are inherently aniso- tropic in thermal expansivity (see Table 2.2); they show zero (or negative) expansion axially on heating, but expand in the transverse (radial) direc- tions.

Levels of thermal stress are of interest for prediction of damage devel- opment. For a fibre content of $\sim$ 40–70%, it can be seen from Fig. 10.8 that stresses of $\sim$ 25–30 MPa are expected. This is approximately the transverse failure stress for typical laminae (see Table 8.2). It follows that cooling through 100 K (from an initially stress-free state) would be likely to cause damage in such a crossply laminate. This is consistent with experimental observations. For example, Fig. 10.9 is a section through a carbon fibre crossply laminate after cycling between 50 °C and -50 °C, showing a crack in the transverse ply. The progressive development of such damage is illustrated by Fig. 10.10, which is a series of X-ray radio- graphs of the same material, taken after different numbers of thermal cycles. The progressive development of cracks parallel to the fibres in each ply shows that the stresses generated during each cooling cycle were close to the critical level for cracking. Such damage is obviously of concern. It may be noted that the requirement to avoid such damage is not that the laminae should have a particular transverse strength level, but rather that they should be able to sustain a particular transverse strain without damage. There is thus interest in the development of poly- mer formulations and manufacturing methods giving high transverse

Fig. 10.9 Optical micrograph (Jennings 1990) of microcracking in a carbon fibre/bismaleimide crossply laminate after thermal cycling.

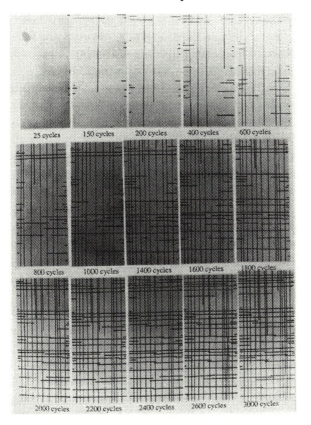

Fig. 10.10   Series of X-ray radiographs (Jennings 1990) showing the development of microcracks parallel to the fibres in a carbon fibre/bismaleimide crossply laminate after thermal cycling between −50 °C and + 50 °C.

strains to failure, particularly for components subject to large temperature fluctuations.

10.2 Creep

10.2.1 Basics of matrix and fibre behaviour

Creep is the term used to describe progressive deformation of a material when subjected to a constant load. This is often undesirable in engineering situations, since in time large strains can develop in a component. Most materials start to exhibit significant creep when appreciable loads (say, > 10 MPa) are imposed at temperatures greater than about 0.4–0.5 $T_{m.p.}$ (i.e. 40–50% of their melting temperature in K). This is essentially due to diffusional processes becoming activated, allowing molecular re-arrangements and associated straining. Thermosetting resins do not have well-defined melting temperatures, but they tend to become chemically and mechanically degraded when subjected to relatively modest temperature increases ($\sim$ 100–200 K) above ambient. They are fairly resistant to creep at room temperature. Thermoplastic polymers, on the other hand, have well-defined glass transition (amorphous) and/or melting (crystalline) temperatures, above which they become (viscous) liquids. These temperatures are in most cases relatively low ($\sim$ 400–600 K), so that they are prone to creep even at room temperature ($\sim$ 300 K). Metals in common use as structural materials range in melting point from $\sim$ 600 K (lead) to $\sim$ 1950 K (titanium). Hence, while most metals show little creep at room temperature, they virtually all become susceptible in the temperature range up to $\sim$ 900 K ($\sim$ 600 °C), which is required for many technological applications.

Typically, the deformation history of a specimen during creep has the form shown in Fig. 10.11. The details of the behaviour of various types of material have been summarised by Frost and Ashby (1982). Broadly, primary creep represents the setting up of some kind of microstructural balance, which is then maintained during the quasi-steady state of secondary creep, before breakdown begins as the tertiary regime is entered. For example, this balance might be one in which dislocations are surmounting obstacles at a rate dictated by diffusional processes and the breakdown might represent irreparable microstructural damage such as

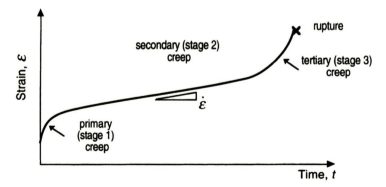

Fig. 10.11   Schematic depiction of a typical strain history during creep deformation under constant load, showing the steady-state creep rate.

internal void formation. Interest commonly focuses on the strain rate during secondary creep, although properties such as the creep rupture strain may also be of concern.

One of the major attractions of composite materials, particularly for thermoplastic and metallic matrices, is that the addition of certain reinforcements can produce dramatic improvements in creep resistance. Glass typically softens at about 900–1000 K (see Table 2.3), but is resistant to creep up to about 550 K. Furthermore, although polymeric fibres (e.g. Kevlar^{TM}) do not have good high-temperature properties, ceramics such as alumina ($T_m \sim 2300$ K) show little creep below 1100 K, while SiC and carbon are in general highly creep resistant to over 1200 K, depending on the details of their microstructure. In most creep situations with composite materials, such fibres can be assumed to behave elastically, i.e. to experience no creep. The creep behaviour of the composite as a whole depends on load partitioning and constraint, which is in turn dependent on geometrical factors and, in some cases, on the nature of the interface.

10.2.2 Axial creep of long-fibre composites

Provided that the fibres remain elastic, treatment of this case is straightforward. The initial strain, generated immediately the composite is loaded, is the applied stress over the Young's modulus of the composite, given by a rule of mixtures (§4.1)

$$\epsilon_0 = \frac{\sigma}{f\, E_f + (1-f)E_m} \qquad (10.13)$$

As creep occurs in the matrix, the applied stress is progressively transferred to the fibres. As the fibres strain elastically, the stress in them increases. The limit comes when the fibres carry all of the applied load. At this point, the strain of the fibres, and hence of the composite, is given by

$$\epsilon_\infty = \frac{\sigma}{f\, E_f} \qquad (10.14)$$

The strain approaches this value asymptotically, since the rate of creep of the matrix falls off as the stress it carries decreases, and a steady state is never set up. From an engineering point of view, this situation is attractive, since the creep strain can never exceed a pre-determined level (which in general is small), assuming that the fibres do not break. Experimental results are consistent with this simple analysis. For example, the data in

Fig. 10.12, from Endo *et al.* (1991), show the strain in an Al/40%SiC composite slowly approaching the calculated limit.

It is possible to predict the rate at which the limiting strain is approached, from a knowledge of the creep characteristics of the matrix. McLean (1983) has shown that, for a matrix which exhibits power-law creep in the steady state $(d\epsilon/dt = A\sigma^n)$, the creep rate of the composite obeys the following equation

$$\frac{d\epsilon}{dt} = \frac{A\sigma^n \left(1 - \dfrac{\epsilon}{\epsilon_\infty}\right)^n}{\left[1 + \dfrac{f E_f}{(1-f)E_m}\right](1-f)^n} \tag{10.15}$$

This equation can be integrated to give the strain as a function of time. In general, this gives fairly good agreement with experiment. In practice, however, the situation may be complicated by factors such as the matrix in the composite differing in some way from the corresponding

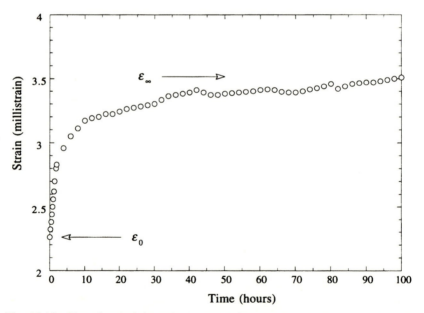

Fig. 10.12 Experimental data (Endo *et al.* 1991) showing the strain as a function of time, for an Al/40 vol. % SiC ('Nicalon^{TM}') long-fibre composite, loaded parallel to the fibres. The applied stress was 266 MPa and the temperature was 673 K. Also shown are the instantaneous and limiting strains, calculated from Eqns (10.13) and (10.14) using stiffness values given in Tables 2.2 and 2.5.

unreinforced material (for example, having a higher dislocation density in the case of metals). The presence of thermal residual stresses (§10.1) may also have an effect, although these tend to become small if the composite is held at a (constant) elevated temperature.

10.2.3 Transverse creep and discontinuously reinforced composites

The situation is very different if the composite is loaded transversely, or if the reinforcement is discontinuous (short fibres or particles). In this case, the composite is able to deform progressively and steady-state creep is often established. Creep rates are strongly influenced by the creep characteristics of the matrix. However, the extent to which the fibres relieve the matrix of load is important. For example, in a short-fibre composite under axial load, the load partitioning depends strongly on the fibre aspect ratio. This is illustrated by Fig. 10.13, which shows the volume-averaged stress in matrix and fibre for elastic loading of a polyester/50% glass composite, as a function of fibre aspect ratio. In this case, the stress in the matrix is reduced by a factor of about 5 as the fibre aspect ratio rises from 1 to 10.

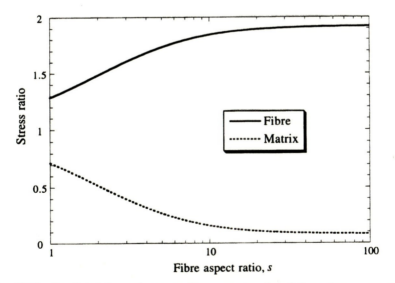

Fig. 10.13 Predicted dependence on fibre aspect ratio of the volume-averaged stress in each constituent, as a ratio to the applied stress, for axial loading of polyester/50% glass fibre composites, obtained using the Eshelby model.

Most experimental data for creep of short-fibre composites under axial load confirm a strong sensitivity to fibre aspect ratio. An example is shown in Fig. 10.14, which gives minimum (steady-state) creep rates for an aluminium alloy reinforced with SiC particles (aspect ratio ~ 1) or with SiC whiskers (aspect ratio ~ 5). The creep rates of the whisker material are lower by about two orders of magnitude, at any given stress level. In many cases, however, such sensitivity to fibre aspect ratio is not just a consequence of the degree of elastic load transfer. During creep, the stress in the matrix is also dependent on the degree to which ***stress relaxation*** occurs. This term is used to describe the unloading of the matrix by diffusive processes which allow it to change shape. These effects occur more readily with low aspect ratio fibres, since diffusion distances are shorter.

Another factor which is sometimes important in creep of discontinuously reinforced composites is the effect of thermal residual stress. When the temperature is changed, there is a change in the level of thermal stress in the matrix (§10.1). This may augment or reduce the stresses from the applied load and hence reduce or enhance the creep rate. Anomalously high values of the activation energy and the stress exponent for creep, often observed for MMCs, have been explained on this basis (Nardone

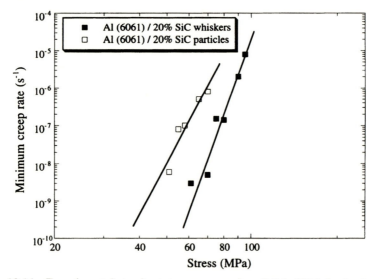

Fig. 10.14 Experimental steady-state creep rate data (Nieh 1984) for isothermal creep at 561 K of Al alloy (6061) reinforced with 20 vol. % of SiC particles or SiC whiskers.

and Tien 1986, Jarry *et al.* 1987). Furthermore, while thermal residual stresses tend to become relaxed and hence are often of minor significance during isothermal creep, they can be of prime importance during thermal cycling, when they are continuously regenerated. Under these circumstances, the creep behaviour can differ substantially from that expected on the basis of the dependence of creep rate on temperature observed during isothermal testing.

The effects outlined above are illustrated by the plots in Fig. 10.15, which give strain history data obtained during thermal cycling creep of short-fibre reinforced aluminium. The predicted plots were obtained by using the Eshelby model (§6.2) to predict the stresses in the matrix from the applied load and as a consequence of the changes in temperature. Instantaneous strain rates were then obtained from the known dependence of matrix creep rate on stress and temperature. Also shown, for the Al/10%Al$_2$O$_3$ composite, are the measured and modelled isothermal creep rates of the composite at the ***diffusional mean temperature*** of the thermal cycle (Wu and Sherby 1984). This is the mean temperature of the imposed cycle, weighted for the temperature-dependence of the creep rate. It can be seen that the average strain rate during cycling is considerably greater than the isothermal rate, reflecting the influence of the thermal residual stresses. It is a cause for concern that such thermal cycling can accelerate the creep of MMCs, since there are many technological applications in which such conditions are imposed.

Finally, the interface sometimes plays an important role during creep of discontinuously reinforced materials. An example is provided by the transverse loading of long-fibre reinforced titanium alloys, which has been examined in some detail (Jansson *et al.* 1994, Newaz and Majumdar 1994). Most of this material is produced currently using SiC monofilaments with a graphitic surface coating. The coating protects the fibre from handling damage and chemical attack during fabrication, but does not allow strong chemical bonding between fibre and matrix. At ambient temperature, fibre/matrix cohesion remains good as a result of the radial compressive thermal residual stress (§10.1). At elevated temperature (> 500 °C), however, the radial stress tends to become tensile, allowing the interface to be opened up by modest applied loads. When the creep rate is also enhanced by thermal cycling, large voids can develop at the interface. This is illustrated by Fig. 10.16, which shows a micrograph of a Ti composite after transverse loading under thermal cycling conditions. Such large interfacial voids lead to high (stage 3) creep rates and rapid onset of creep rupture.

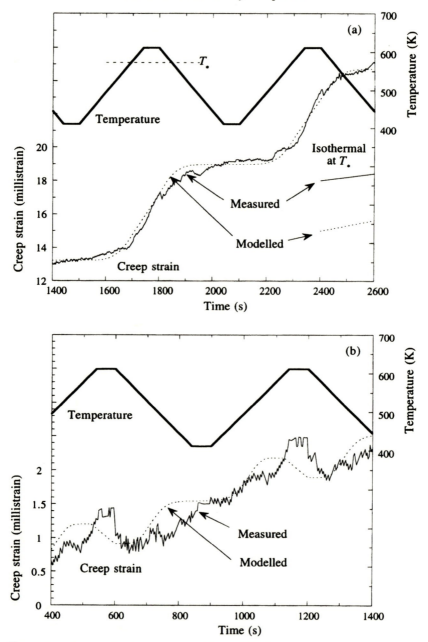

Fig. 10.15   Comparison (Furness and Clyne 1991) between measured and modelled in-cycle creep strain history for (a) Al/10% Saffil^{TM} (short Al$_2$O$_3$ fibre) and (b) Al/20% Saffil^{TM}, under 20 MPa applied axial load. Data are also shown in (a) for the corresponding isothermal creep at the diffusional mean temperature.

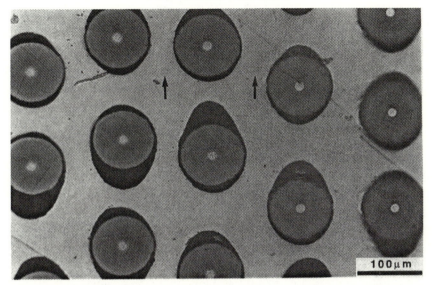

Fig. 10.16 SEM micrograph (Clyne *et al.* 1995) of a polished section near the fracture surface of a Ti–6Al–4V/30% SiC composite, after rupture under transverse loading at a stress of 25 MPa with thermal cycling (between 400 °C and 700 °C, at 100 °C per minute). The direction of loading is indicated by the arrows.

10.3 Thermal conduction

High thermal conductivity is useful in improving the resistance of materials to thermal shock and avoiding the development of 'hot spots' during localised heating. In other situations, a low thermal conductivity can be beneficial in providing thermal insulation. Before considering conductivity levels in fibres, matrices and composites, it is useful to review the mechanisms of thermal conduction.

10.3.1 Heat transfer mechanisms

Heat flows within a material by the transmission of phonons (lattice vibrations) and free electrons (if present). Both of these *carriers* have a certain mean free path λ between collisions (energy exchange events) and an average velocity v. The thermal conductivity K is related to these parameters by a simple equation derived from kinetic theory

$$K = \frac{1}{3}cv\lambda \qquad (10.16)$$

where c is the volume specific heat of the carrier concerned. Both metals (electron-dominated) and non-metals (phonons only) have zero conductivity at a temperature of 0 K (where c becomes zero in both cases), followed by a sharp rise to a peak and then a gradual fall as the temperature is progressively increased. The rise reflects the increase in c towards a plateau value ($\sim 3\,Nk$, where N is Avogadro's number and k is the Boltzmann constant, with all the vibrational modes of all the molecules active). The fall is caused by the decreasing λ as the greater amplitude of lattice vibration causes more scattering of the carriers. The maximum (low-temperature) value of λ is dictated by atomic-scale defects for electrons and by the physical dimensions of the specimen for phonons.

The average carrier velocity is insensitive to temperature in both cases. The phonon velocity (speed of sound) is high in light, stiff materials. The mean free path of a phonon is structure-sensitive and can be very large in pure specimens of high perfection and large grain size. Single crystals of materials like diamond and SiC can therefore have very high thermal conductivities. Some recently developed pitch-based carbon fibres also have very high thermal conductivities (Kowalski 1987). With the exception of such cases, metals tend to have the highest conductivities, because electrons usually have a much larger mean free path than phonons. This is substantially reduced by the presence of solute atoms and various defects which cause electron scattering. Polymers have no free electrons and low stiffnesses (low phonon velocity), so that conductivities are low. These trends are all apparent in the data shown in Table 10.1.

10.3.2 Conductivity of composites

The basic equation of heat flow may be written

$$q = -KT' \tag{10.17}$$

where q is the heat flux ($\mathrm{W\,m^{-2}}$) arising from a thermal gradient T' ($\mathrm{K\,m^{-1}}$) in a material of thermal conductivity K ($\mathrm{W\,m^{-1}\,K^{-1}}$). It is important to distinguish K from the thermal diffusivity $a\ (= K/c)$, which is the parameter determining the rate at which a material approaches thermal equilibrium. This appears in the unsteady diffusion equation

$$\frac{\partial T}{\partial t} = aT'' \tag{10.18}$$

Table 10.1 *Thermal conductivity data for a range of materials*

Material	K (W m^{-1} K^{-1})	
	at 300 K	at 900 K
Diamond	600	—
Graphite (parallel to *c*-axis)	355	—
Graphite (normal to *c*-axis)	89	—
Ag	425	325
Cu	400	340
Cu–2%Ag	~ 390	~ 340
Cu–2%Be	~ 130	—
Cu–40%Ni	~ 20	~ 40
Al	~ 220	~ 180
Ti	~ 18	~ 18
Ti–6Al–4V	~ 8	~ 12
SiC (single crystal)	~ 100	~ 70
SiC (polycrystal)	~ 10–50	~ 5–30
Al$_2$O$_3$ (single crystal)	~ 100	~ 30
Al$_2$O$_3$ (polycrystal)	~ 5–30	~ 3–10
Epoxy resin	0.2–0.5	—
Polyester resin	0.2–0.24	—
Nylon	0.2–0.25	—

The conductivity of a composite can be predicted provided suitable assumptions are made about the flow of heat through the constituents, i.e. the shape of the isotherms. Several reviews are available covering the effective conductivity of composites (Nielsen 1974, Hale 1976, Hatta and Taya 1986). For heat flow in long-fibre composites, the slab model (§4.1) can be used. Axial and transverse heat flow are represented schematically in Fig. 10.17. For the axial case, the thermal gradient (spacing between the isotherms) is the same in each constituent. The total heat flux is given by the sum of the flows through fibre and matrix

$$q_{1c} = f q_{1f} + (1 - f) q_{1m} \qquad (10.19)$$

These can be written in terms of conductivities and thermal gradients

$$K_{1c} T' = f K_f T' + (1 - f) K_m T'$$

so that the composite conductivity is given by a simple rule of mixtures

$$K_{1c} = f K_f + (1 - f) K_m \qquad (10.20)$$

This expression should give a reliable value for the axial conductivity.

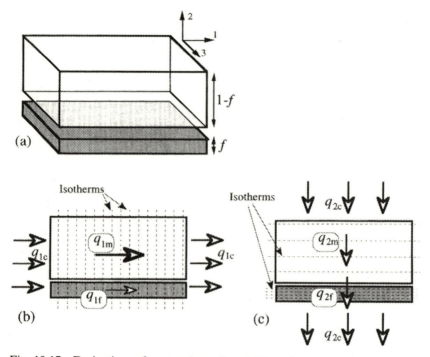

Fig. 10.17 Derivation of expressions for axial and transverse conductivity. Schematic depiction of (a) the slab model for long-fibre composites, (b) axial heat flow and (c) transverse heat flow.

The transverse conductivity, based on the slab model, is obtained by equating the heat fluxes through the two constituents (Fig. 10.17(c))

$$q_{2c} = K_{2c}T'_c = q_{2f} = K_f T'_f = q_{2m} = K_m T'_m \qquad (10.21)$$

The thermal gradients in the two constituents are related to the overall average value by the expression

$$T'_c = f\, T'_f + (1-f)T'_m$$

leading to the following expression for the transverse conductivity

$$K_{2c} = \left(\frac{f}{K_f} + \frac{1-f}{K_m}\right)^{-1} \qquad (10.22)$$

This is analogous to the expression for the transverse stiffness, derived in §4.2. As with that derivation, the assumption that the two constituents lie 'in series' with each other leads to this expression being of poor accuracy.

In fact, it represents a lower bound. It becomes very unreliable when the fibres have a low conductivity; for insulating fibres, the composite is predicted to have zero conductivity even when the fibre volume fraction is small.

More reliable treatments are available for the transverse conductivity. For example, the Eshelby method (§6.2) can be adapted to predict the conductivity of composites with reinforcement having any aspect ratio (Clyne and Withers 1993). The tensor equation obtained is

$$K_c = [K_m^{-1} + f\{(K_m - K_f)[S - f(S - I)] - K_m\}^{-1}(K_f - K_m)K_m]^{-1}$$

(10.23)

which is closely analogous to Eqn (6.34) for stiffness. Hatta and Taya (1986) have shown that, for the transverse conductivity of a long-fibre composite, this reduces to the expression

$$K_{2c} = K_m + \frac{K_m(K_f - K_m)f}{K_m + (1 - f)(K_f - K_m)/2}$$

(10.24)

Predictions obtained from these equations are shown in Fig. 10.18. The rule of mixtures expression gives a reliable prediction for the axial conductivity of long-fibre composites, but it can be seen that the slab model expression for transverse conductivity (Eqn (10.22)), which is completely unreliable for insulating fibres, also gives a large error when the fibres have high conductivity. For highly conducting fibres, the aspect ratio has little effect on the transverse conductivity and Eqn (10.24) provides a good approximation for most cases. The axial conductivity, on the other hand, is more sensitive to aspect ratio and Eqn (10.20) gives an overestimate for short-fibre and particulate composites.

An example is given in Fig. 10.19 of how such predictions can be used in exploring performance characteristics of different composite systems. This shows a map (Ashby 1993) of thermal expansivity against thermal conductivity, for aluminium reinforced with silicon carbide or boron nitride. The ratio K/α can be taken as a merit index for minimisation of thermal distortion during heating or cooling. On this plot, the merit index is high in the bottom right and low in the top left. The data shown therefore indicate that the resistance of aluminium to thermal distortion is improved by the incorporation of silicon carbide, but is impaired if boron nitride is added.

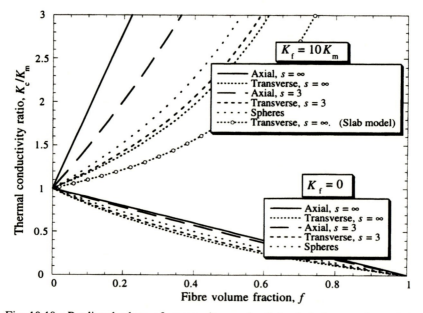

Fig. 10.18 Predicted plots of composite conductivity (relative to that of the matrix) as a function of fibre content, for insulating ($K_f = 0$) and highly conducting ($K_f = 10K_m$) reinforcements having various aspect ratios. These curves were obtained using the Eshelby method (Eqn (10.23)), except for the transverse conductivity of a long-fibre composite with highly conducting fibres, in which case the prediction of Eqn (10.22), obtained from the slab model, is also shown.

10.3.3 Interfacial thermal resistance

The above calculations are based on the assumption that there is perfect thermal contact between fibre and matrix. In practice, this may be impaired by the presence of an interfacial layer of some sort, or by voids or cracks in the vicinity of the interface. Furthermore, even in the absence of such barriers to heat flow, there may be some loss of heat transfer efficiency across the interface if the carriers are different in the two constituents, as with metal/ceramic systems. Such a thermal resistance is characterised by an interfacial **heat transfer coefficient** or thermal conductance, h (W m^{-2} K^{-1}), defined as the proportionality constant between the heat flux through the boundary and the temperature drop across it

$$q_i = h \Delta T_i \qquad (10.25)$$

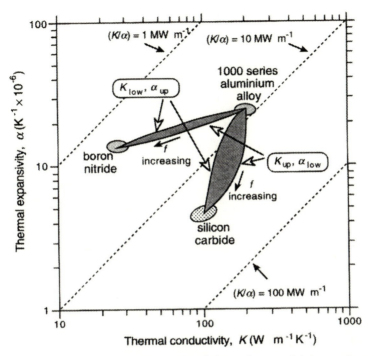

Fig. 10.19 Predicted map (Ashby 1993) of thermal expansivity α against thermal conductivity K for composites made up of either silicon carbide or boron nitride in an aluminium matrix. The diagonal dotted lines represent constant values of a merit index, given by K/α, taken as indicative of the resistance to thermal distortion. The shaded areas, defined by upper and lower bounds of the two parameters, indicate the possible combinations of K and α expected for Al/SiC and Al/BN composites, depending on the volume fraction, shape and orientation distribution of the reinforcement.

Modelling of the thermal conductivity of composites in the presence of such an interfacial resistance has been addressed by several authors (Benveniste and Miloh 1986, Hasselman and Johnson 1987, Fadale and Taya 1991, Hasselman *et al.* 1993). Hasselman and Johnson derived an analytical expression for the transverse conductivity of a long-fibre composite.

$$K_c = K_m \frac{\left[f\left(\dfrac{K_f}{K_m} - \dfrac{K_f}{rh} - 1 \right) + \dfrac{K_f}{K_m} + \dfrac{K_f}{rh} + 1 \right]}{\left[f\left(1 - \dfrac{K_f}{K_m} + \dfrac{K_f}{rh} \right) + \dfrac{K_f}{K_m} + \dfrac{K_f}{rh} + 1 \right]} \qquad (10.26)$$

where r is the radius of the fibres. The scale of the structure is now relevant because it determines how frequently interfaces are encountered by the heat flow. This scale effect is represented by the dimensionless ratio K_f/rh.

Equation (10.26) can be used in conjunction with experimental conductivity data to estimate the interfacial conductance of a composite. For example, Fig. 10.20 shows measured conductivities over a range of temperature for a Ti-based long-fibre composite, plotted as a ratio to that of the matrix. Also shown are predictions from Eqn (10.26) corresponding to several h values, using the appropriate fibre radius ($\sim 50\,\mu$m). These data are consistent with an h value of around $10^6\,\mathrm{W\,m^{-2}\,K^{-1}}$. Although this figure represents relatively poor thermal contact for a fibre/matrix interface, it can be seen that in this system the overall conductivity would not have been much greater had the interface been perfect ($h = \infty$). This illustrates the size effect; the large diameter of the SiC monofilaments means that the magnitude of

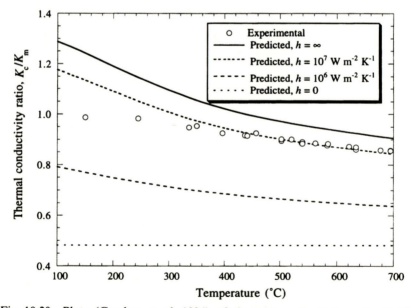

Fig. 10.20 Plots (Gordon *et al.* 1994) of thermal conductivity ratio K_c/K_m against temperature, for a Ti–6Al–4V alloy reinforced with 35% of SiC monofilament, with heat flow transverse to the fibre axis. Both experimental data and predictions from Eqn (10.26), based on several values for h (in $\mathrm{W\,m^{-2}\,K^{-1}}$), are shown.

interfacial resistance is not of much significance (unless it becomes very large indeed). It may also be noted that, although data in Table 10.1 suggest that there is scope for raising the conductivity of Ti–6Al–4V by addition of SiC, the monofilament in this composite has a relatively low conductivity ($\sim 10\,\mathrm{W\,m^{-1}\,K^{-1}}$), primarily as a consequence of its very fine grain size.

Two additional points can be illustrated by some data for particulate-reinforced Ti. For a volume fraction f of spherical particles (having a radius r) in a relatively dilute composite, an analytical expression given by Hasselman and Johnson (1987) can be used to predict the conductivity of the composite

$$K_c = K_m \frac{\left[2f\left(\dfrac{K_p}{K_m} - \dfrac{K_p}{rh} - 1\right) + \dfrac{K_p}{K_m} + 2\dfrac{K_p}{rh} + 2\right]}{\left[f\left(1 - \dfrac{K_p}{K_m} + \dfrac{K_p}{rh}\right) + \dfrac{K_p}{K_m} + 2\dfrac{K_p}{rh} + 2\right]} \tag{10.27}$$

where K_p is the conductivity of the particles. In Fig. 10.21(a), experimental conductivity data for a $\mathrm{Ti}/10\%\,\mathrm{SiC_p}$ composite have been plotted as ratios to that of the matrix, over the range of temperature. Also shown are predictions from Eqn (10.27) corresponding to several h values, using the experimental conductivity values for matrix and reinforcement at the appropriate temperatures, and a particle radius of $10\,\mu\mathrm{m}$. These data are again consistent with an h value of around $10^6\,\mathrm{W\,m^{-2}\,K^{-1}}$. In this case, however, a consequence of the smaller size of the reinforcement, compared with the fibre composite data shown in Fig. 10.20, is that the overall conductivity is considerably lower than would be the case with perfect interfacial contact. Indeed, the conductivity is quite close to that expected with an insulating interface ($h = 0$).

From the data in Fig. 10.21(b), it is evident that the interfacial characteristics are different for the $\mathrm{Ti}/\mathrm{TiB_2}$ particulate MMC. These plots show that the interfacial conductance is higher in this system, by a factor of about 10, giving conductivities close to those expected with perfect interfacial contact. The difference between the two systems has been attributed (Gordon *et al.* 1994) to the thicker interfacial reaction layer in the Ti/SiC composites and the fact that the reaction involves a large volume reduction, tending to cause interfacial cracks.

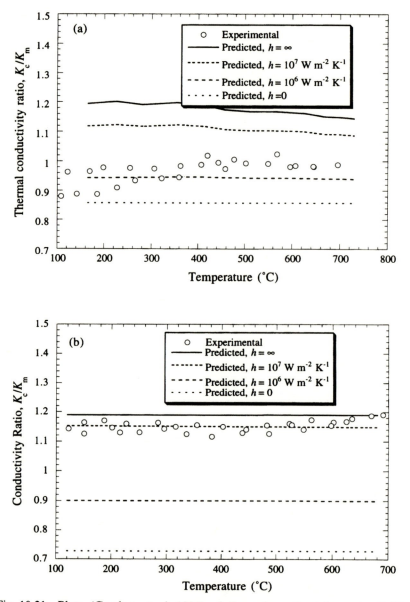

Fig. 10.21 Plots (Gordon *et al.* 1994) of thermal conductivity ratio K_c/K_m against temperature for two particle-reinforced composites, showing both experimental data and predictions from Eqn (10.27), based on several values for h (in $W\,m^{-2}\,K^{-1}$) and using measured conductivities for both matrix and reinforcement. Plots are shown for (a) Ti/10% SiC and (b) Ti/10% TiB$_2$.

References and further reading

Ashby, M. F. (1993) Criteria for selecting the components of composites, *Acta Metall. Mater.*, **41** 1313–35

Benveniste, Y. and Miloh, T. (1986) The effective thermal conductivity of composites with imperfect thermal contact at constituent interfaces, *Int. J. Eng. Sci.*, **24** 1537–52

Bowles, D. E. and Tompkins, S. S. (1989) Predictions of coefficients of thermal expansion for unidirectional composites. *J. Comp. Mat.*, **23** 370–88

Clyne, T. W. and Withers, P. J. (1993) *An Introduction to Metal Matrix Composites.* Cambridge University Press: Cambridge

Clyne, T. W., Feillard, P. and Kalton, A. F. (1995) Interfacial mechanics and macroscopic failure in titanium-based composites, in *Life Prediction Methodology for Titanium Matrix Composites.* W. S. Johnson (ed.) ASTM STP

Endo, T., Chang, M., Matsuda, N. and Matsuura, K. (1991) Creep behavior of SiC/Al composite at elevated temperature, in *Metal Matrix Composites – Processing, Microstructure and Properties.* N. Hansen *et al.* (eds.) Risø Nat. Lab.: Roskilde, Denmark pp. 323–8

Fadale, T. D. and Taya, M. (1991) Effective thermal conductivity of composites with fibre–matrix debonding, *J. Mater. Sci. Letts.*, **10** 682–4

Frost, H. J. and Ashby, M. F. (1982) *Deformation Mechanism Maps – the Plasticity and Creep of Metals and Ceramics.* Pergamon: Oxford

Furness, J. A. G. and Clyne, T. W. (1991) Thermal cycling creep of short fibre MMCs – measurement and modelling of the strain cycle, in *Metal Matrix Composites – Processing, Microstructure and Properties.* N. Hansen *et al.* (eds.) Risø Nat. Lab.: Roskilde, Denmark pp. 349–54

Gordon, F. H., Turner, S. P., Taylor, R. and Clyne, T. W. (1994) The effect of the interface on the thermal conductivity of Ti-based composites, *Composites*, **25** 583–92

Hale, D. K. (1976) The physical properties of composite materials, *J. Mater. Sci.*, **11** 2105–41

Hasselman, D. P. H. and Johnson, L. F. (1987) Effective thermal conductivity of composites with interfacial thermal barrier resistance, *J. Comp. Mat.*, **21** 508–15

Hasselman, D. P. H., Donaldson, K. Y. and Thomas, J. R. (1993) Effective thermal conductivity of uniaxial composite with cylindrically orthotropic carbon fibres and interfacial barrier resistance, *J. Comp. Mat.*, **27** 637–44

Hatta, H. and Taya, M. (1986) Thermal conductivity of coated filler composites, *J. Appl. Phys.*, **59** 1851–60

Jansson, S., Dalbello, D. J. and Leckie, F. A. (1994) Transverse and cyclic thermal loading of the fiber reinforced metal matrix composite SCS6/Ti–15-3, *Acta Mater. Metall.*, **42** 4015–24

Jarry, P., Loué, W. and Bouvaist, J. (1987) Rheological behaviour of SiC/Al composites, in *Proc. ICCM6.* F. L. Matthews *et al.* (eds.) Elsevier: London pp. 2.350–2.361

Jennings, T. M. (1990) *Thermal Fatigue of Carbon Fibre–Bismaleimide Composites.* Ph.D thesis, University of Cambridge

Kowalski, I. M. (1987) New high-performance domestically produced carbon fibers, *SAMPE J.*, **32** 953–63

Kural, M. H. and Min, B. K. (1984) The effects of matrix plasticity on the thermal deformation of continuous fibre graphite/metal composites, *J. Comp. Mat.*, **18** 519–35

Lamé, G. (1852) *Leçons sur la Théorie de l'Élasticité*. Gautiers–Villars: Paris

Lee, J. K., Earmme, Y. Y., Aaronson, H. I. and Russel, K. C. (1980) Plastic relaxation of the transformation strain energy of a misfitting spherical precipitate: ideal plastic behaviour, *Metall. Trans.*, **11A** 1837–47

Masutti, D., Lentz, J. P. and Delannay, F. (1990) Measurement of internal stresses and of the temperature dependence of the matrix yield stress in metal matrix composites from thermal expansion curves, *J. Mat. Sci. Letts.*, **9** 340–2

Mclean, M. (1983) *Directionally Solidified Materials for High-Temperature Service*. Metals Soc.: London

Nardone, V. C. and Tien, J. K. (1986) On the creep rate dependence of particle strengthened alloys, *Scripta Met.*, **20** 797–802

Newaz, G. M. and Majumdar, B. S. (1994) A comparison of mechanical response of MMC at room and elevated temperature, *Comp. Sci. & Tech.*, **50** 85–90

Nieh, T. G. (1984) Creep rupture of a silicon carbide reinforced aluminium composite, *Metall. Trans.*, **15A** 139–46

Nielson, L. E. (1974) The thermal and electrical conductivity of two-phase systems, *Ind. Eng. Chem. Fundam.*, **13** 17–20

Schapery, R. A. (1968) Thermal expansion coefficients of composite materials based on energy principles, *J. Comp. Mat.*, **2** 380–404

Withers, P. J., Jensen, D. J., Lilholt, H. and Stobbs, W. M. (1987) The evaluation of internal stresses in a short fibre MMC, in *Proc. ICCM6*. F. L. Matthews *et al.* (eds.) Elsevier: London pp. 2.255–2.264

Wolff, B. K., Min, B. K. and Kural, M. H. (1985) Thermal cycling of a unidirectional graphite–magnesium composite, *J. Mat. Sci.*, **20** 1141–9

Wu, M. Y. and Sherby, O. D. (1984) Superplasticity in a silicon carbide reinforced aluminium alloy, *Scripta Met.*, **18** 773–6

11

Fabrication

An important aspect of composite materials concerns the technology by which they are produced. Depending on the nature of matrix and fibre, and the required architecture of the fibre distribution, production at reasonable cost and with suitable microstructural quality can be a challenging problem. In most cases, manufacture of the final component and production of the composite material are carried out at the same time. This gives scope for optimal fibre placement, but also demands that the mechanical requirements of the application be well understood and that the processing route be tailored accordingly. Fabrication procedures for most polymer composites are commercially and technically mature, while most of those being applied to metal and ceramic composites are still under development. In many such cases, commercial exploitation will be dependent on improved fabrication efficiency.

11.1 Polymer composites

There are many commercial processes for the manufacture of PMC components. These may be sub-divided in a variety of ways, but broadly speaking there are three main approaches to the manufacture of fibre-reinforced thermosetting resins and two distinct production methods for thermoplastic composites. These are briefly covered below under separate headings. A simple overview of the starting materials and approaches adopted to their incorporation into components is given in Fig. 11.1. In most cases, the main microstructural objectives are to ensure that the fibres are well wetted, uniformly distributed and

271

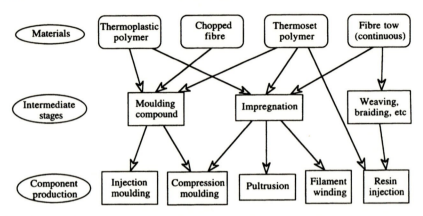

Fig. 11.1 Schematic overview of the approaches employed in fabrication of
polymer matrix composites.

correctly aligned. Practical considerations relating to capital cost, speed
of production and component size and shape capability are often of
paramount importance.

11.1.1 Liquid resin impregnation routes

In the basic processing route, low-viscosity resin is impregnated into
arrays of dry fibre. This can be achieved in several ways. The simplest
is a wet lay-up procedure in which the fibres, usually in the form of a mat
on a polished *former* or mould, are impregnated with resin by rolling or
spraying. The resin and curing agent are mixed immediately prior to
application. Curing usually takes place at ambient temperature. Recent
developments include co-spraying of liquids so that mixing occurs as the
liquids are introduced to the fibres (Gotch 1994). The main advantages of
the process lies in its versatility. Virtually any shape can be produced and
capital costs are limited to that of the mould. The technique has tradi-
tionally been used in a variety of small-scale operations, such as car body
repair, commonly with glass fibres in the form of woven roving or
chopped strand mat (see Figs 3.6 and 3.7, respectively). Larger-scale
operation is, however, increasingly common: for example, the technique
is widely used in the boat-building industry for production of relatively
large craft such as minesweepers of up to about 50 m in length (Smith
1990). The main disadvantages lie in the difficulty of ensuring complete
impregnation and the labour-intensive nature of much of the work.

A process more suited to automation, although limited to certain component shapes, is *filament winding*. Fibre tows, i.e. bundles of fibres, are drawn through a bath of resin, before being wound onto a mandrel or former of the required shape. The equipment comprises (a) a creel stand, from which the fibre tows are fed under the required tension from a set of reels, (b) a bath of resin, through which the fibre tows pass via a set of guides, (c) a delivery eye, through which the fibres emerge, the position of which is controlled by a mechanical system and (d) a rotating mandrel, onto which the fibre tows are drawn. The key parameters are the fibre tension, the resin take-up efficiency and the winding geometry (Middleton 1994). There are a number of different designs for the movement of the delivery eye. One is shown in Fig. 11.2, incorporating a gantry for the traversing motion.

In most cases, the eye motion and mandrel rotation systems are computer-controlled. Component shapes can be fairly complex, although they often exhibit a high degree of symmetry. There are also some limitations on the paths that the fibres take over the surface of the component. On any curved surface, there will be a tendency for the fibres to follow a *geodesic* path – i.e. the shortest one. This can cause problems with some shapes, since it may be difficult to ensure that fibres

Fig. 11.2   Photograph of a gantry-type filament winding machine (Middleton 1994).

cover some parts of the surface, or lie in certain orientations. It is, however, possible to ensure that fibres follow certain non-geodesic paths, provided the delivery eye is moved appropriately and there is sufficient friction between the tow and the underlying surface (Wells and McAnulty 1987). Filament winding is often used to produce high-performance components and is obviously well suited to simple shapes such as tubes.

Pultrusion is a process which is similar in some respects to filament winding. This also involves fibre tows being passed through a resin bath. In this case, the impregnated tows are then fed into a heated, tubular die, in which they become consolidated and the resin is cured. The die may have a relatively complex sectional shape. The composite is pulled from the die by a frictional extraction system. The process generates stock material, rather than finished components, and the product is similar in that respect to plastic and metallic material produced by conventional extrusion.

Another impregnation route involves injection of resin into a mould in which the fibres are placed in position. The resin is fed in under gravity or external pressure. There are several different types of machine. A commonly used process is that termed *resin transfer moulding* (RTM), in which the fibres are enclosed in a die and pre-catalysed resin, having a low viscosity, is injected into the mould at relatively low pressure. Cure occurs within the mould, often assisted by heating. The mould is usually of metal, which gives good heat transfer and lasts for many moulding operations. A typical set-up is illustrated in Fig. 11.3, which shows the mould for production of a tennis racquet, with a carbon fibre preform already in place. To encourage good infiltration, the mould is sometimes evacuated prior to injection of the resin. Relatively large mouldings, including many body parts for the automobile industry, are made in this way.

11.1.2 Pressurised consolidation of resin pre-pregs

This approach involves production of a *pre-preg*, which is a tape or sheet of fibres impregnated with resin. Pre-preg is manufactured by laying the fibres and resin between sheets of siliconised paper or plastic film, which are pressed or rolled to ensure consolidation and wetting out of the fibres, and then partially cured to produce a flexible aggregate. The process allows excellent alignment of the fibres in unidirectional layers. A component is formed by stacking up the layers of pre-preg in pre-determined directions, consolidating them by pressure, and finally curing by heating

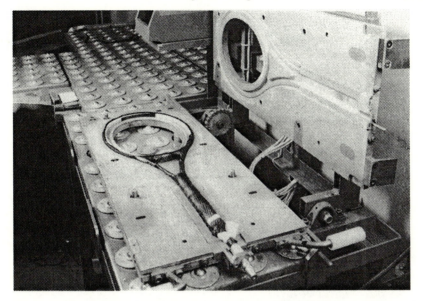

Fig. 11.3   Photograph of equipment for resin transfer moulding production of a carbon fibre reinforced tennis racquet (Blake 1989).

under pressure. The simplest arrangement involves a directional hydraulic press, which is used to apply pressure to a pair of matched mould halves. An alternative approach, *vacuum moulding*, involves enclosing the pre-preg (draped over a former) with a flexible, impermeable membrane. The enclosure is then evacuated, so that atmospheric pressure acts on the membrane and compresses the pre-preg onto the former.

While the above processes are relatively quick and easy to carry out, they suffer from limitations in terms of the quality of the composite material produced and the size and shape of components that can be manufactured. For demanding applications, *autoclave moulding* is often preferred. This is similar to vacuum moulding, but the enclosed composite assembly is placed into a large chamber which can be pressurised, typically up to ~ 10–20 atmospheres. Heating is applied during pressurisation to cure the resin. In many cases, the temperature is first raised to an intermediate level, in order to reduce the viscosity of the resin and ensure that all voids are removed (Jones 1994). Further heating then ensures that the cure is complete. A schematic representation of the changes in temperature, pressure and resin viscosity during autoclave moulding is shown in Fig. 11.4. The process can be applied to large

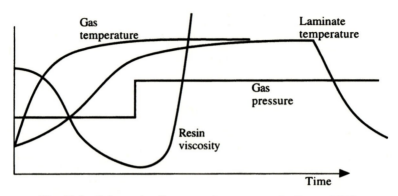

Fig. 11.4   Schematic of an autoclave cure cycle (Jones 1994).

components. This is illustrated by Fig. 11.5(a), which shows a complete aeroplane wing being moved into a large autoclave. The plane concerned is shown in Fig. 11.5(b).

11.1.3 Consolidation of resin moulding compounds

The third approach for thermosetting resins is a variation of the pre-preg approach. The intermediate products are ***sheet moulding compound*** (SMC) and ***dough moulding compound*** (DMC), which are usually based on polymer resins. To make SMC, resin containing thickening agents and particulate fillers such as calcium carbonate, is mixed with chopped fibres to form a slurry. This is then fed between two thermoplastic films and fed over a series of rolls, so that impregnation takes place and a consolidated sheet is produced. The fibres lie mainly in the plane of the sheet and the fibre volume fraction is usually in the range 15 to 40%. Sheets are typically between 3 and 10 mm thick. DMCs are made by mixing together similar constituents to those in SMCs, but the mixing is carried out differently. Usually, a blender generating high shear rates is used for DMCs. This tends to break the fibres up into shorter lengths and also results in a more random orientation distribution.

Both SMC and DMC are hot press moulded in closed moulds to produce the final product and, in the process, they undergo the final cure. SMCs are commonly processed into components by hot press or compression moulding. This is done by removing the thermoplastic films, cutting the sheets to suitable shapes and placing them in a heated mould.

(a)

(b)

Fig. 11.5 (a) Photograph of a large autoclave being used for production of the wing of a Beech Starship aeroplane. (b) Photograph of a Beech Starship in flight (Blake 1989).

Similar operations are carried out with DMCs, although in this case the mixture is sometimes formed into more complex 3-D shapes. A related process is *reaction injection moulding* (Becker 1979), which can be done with or without fibre reinforcement. This also bears some similarity to

resin transfer moulding, but instead of using a pre-catalysed resin which
cures slowly, as with RTM, two components which react together quickly
are mixed, together with some short fibres, and immediately injected into
a mould. In all such processes, the presence of fibres raises the viscosity of
the mixture dramatically and strongly inhibits its flow, even when they
are relatively short, so that the moulding operation may become imprac-
tical unless the fibre content is low.

11.1.4 Injection moulding of thermoplastics

Short-fibre reinforced thermoplastic materials are used to make components by **injection moulding**. This process is very widely applied to
unreinforced thermoplastics. Three polymers for which the process is
very common are polypropylene, nylon and polycarbonate (see Table
2.5). The first two are semi-crystalline with approximately 25–50%
crystallinity and the latter is amorphous. Polymer pellets are fed into
the heated barrel of a chamber containing an Archimedes screw for
transporting the charge. The shear motion helps to homogenise and
melt the polymer, which is then periodically injected into a mould. The
same process is carried out using pellets containing short fibres,
typically 1–5 mm long, intimately mixed and dispersed in the matrix.
However, the comments in §11.1.3 about flow of a mixture becoming
difficult when fibres are present also applies to these operations, and
the volume fraction of fibres is usually no greater than about
10–20%.

Of interest for injection moulded components is the fibre orientation
distribution. For example, it may be possible to arrange for the fibres to
be oriented parallel to the direction of stress in parts of the component
which will be heavily loaded in service. Fibre orientation is controlled by
the nature of the flow field during filling of the mould. In general, fibres
tend to become aligned parallel to the direction in which the material is
becoming elongated (provided the flow is not too turbulent). This is
illustrated by the schematic diagram of the flow pattern during injection
into a simple rectangular mould, and the corresponding micro-radio-
graphs, shown in Fig. 11.6. In this case, the fibres are well aligned in
the outer layers of the moulding, but more randomly oriented towards
the core. By predicting the flow behaviour under different injection conditions for specific components, a degree of control over the final fibre
orientation pattern is often possible.

(a)

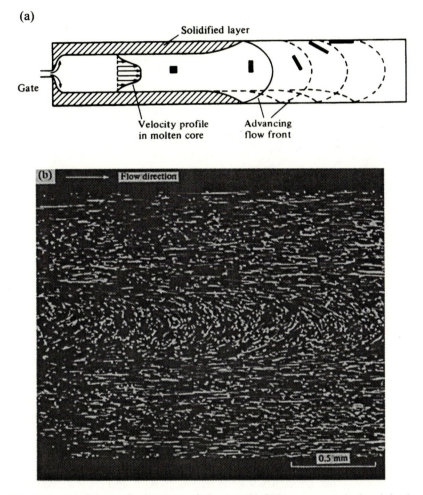

Fig. 11.6 (a) Schematic diagram of the mould filling process during injection moulding, showing the deformation of an initially square fluid element at successive positions of the advancing flow front (Folkes and Russell 1980). (b) Contact microradiograph of the longitudinal section of a polypropylene/15% glass fibre injection moulding. (c) [*overpage*] Transverse section from the same moulding.

11.1.5 Hot press moulding of thermoplastics

Long- or continuous-fibre reinforced thermoplastics are commonly used in *laminated structures*. The first stage in manufacture is the production of a pre-preg by *melt impregnation* of the fibres, which may be in the form of continuous aligned sheets, woven cloth, etc. The pre-preg sheets are then stacked in the required orientations and *hot pressed* to form the final

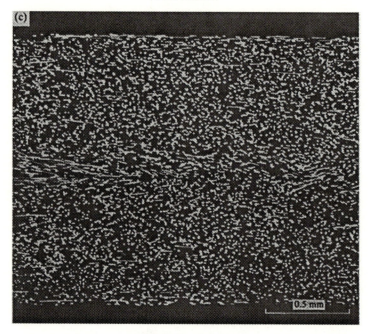

Fig. 11.6   *(cont)*

product. Pre-pregs are available in many materials including polypropylene, polysulphone and polyether-ether-ketone (PEEK). Processing is
carried out in a similar manner to that involved during pressurised consolidation of resin pre-pregs (§11.1.2). For thermoplastics, there is no
requirement for the matrix to be cured and the temperature is usually
set to the minimum necessary for the matrix to melt and flow sufficiently
for consolidation to occur. However, even when fully molten, thermoplastic melts are much more viscous than uncured resins, since thermoplastics are already fully polymerised. Substantial heating and pressure
are therefore necessary, and impregnation distances into fibre arrays are
always kept as short as possible.

11.2 Metal composites

Production of MMCs is commercially less advanced than PMCs.
Industrial exploitation of MMCs is still in its infancy and mature production technologies have yet to emerge. Nevertheless, there has been
considerable research effort into fabrication aspects, partly because the

difficulties and cost of production have been largely responsible for their usage being relatively limited. The diversity of approaches which have been attempted is reflected in the complexity of the overview presented in Fig. 11.7. Several of the production routes represented in this figure are of limited commercial potential, since they are inherently expensive or cumbersome. Some of them may, nevertheless, find industrial use in due course, since there are certain applications for which a high manufacturing cost is justifiable. A basic distinction can be drawn, depending on whether or not liquid metal is involved. Broadly, the liquid metal routes are relatively cheap (apart from the spraying techniques, which have now been largely discarded), while the solid-state methods offer certain advantages at increased cost. In the classification given below, the first three routes involve contact between liquid metal and the ceramic, whereas in the following three there is no such contact.

11.2.1 Squeeze infiltration

The most common pressure-assisted solidification process for MMC production is *squeeze infiltration*. Liquid metal is injected into an assembly of short fibres, usually called a 'preform'. Commonly, the preform is designed with a specific shape to form an integral part of a finished product in the as-cast form (Feest 1988). Preforms are commonly fabricated by sedimentation of short fibres from liquid suspension, often using

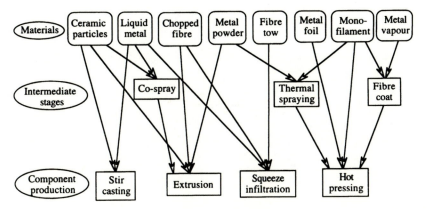

Fig. 11.7 Schematic overview of the approaches employed in fabrication of metal matrix composites.

short alumina fibres, such as 'Saffil^{TM}' (see §2.1.5). The process can also be adapted for production of particulate MMCs (Klier *et al.* 1991). In order for the preform to retain its integrity and shape, a **binder** (usually silica-based) is used.

In most cases the fibres do not act as preferential crystal nucleation sites during melt solidification (Mortensen and Cornie 1987). One consequence of this is that the last liquid to freeze, which is normally solute-enriched, tends to be located around the fibres. An example of this is shown in Fig. 11.8. Such prolonged fibre/melt contact, under high hydrostatic pressure and with solute enrichment, tends to favour formation of a strong interfacial bond. Furthermore, oxide films cannot form because of the limited oxygen availability (Clyne and Mason 1987). This probably contributes to the high interfacial bond strengths commonly observed.

11.2.2 Stir casting

This technique involves stirring liquid metal with solid ceramic particles and then allowing the mixture to solidify. This can be done using fairly

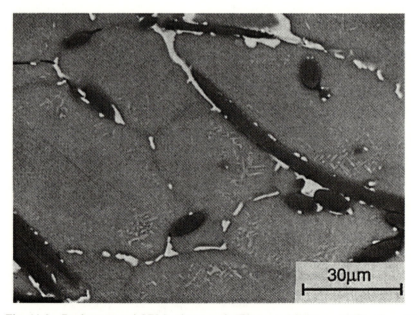

Fig. 11.8   Back-scattered SEM micrograph (Clyne and Withers 1993) of an Al–4.5 wt% Cu/Saffil^{TM} composite, showing copper-rich (light) regions around the fibres.

conventional processing equipment and is carried out on a continuous or semi-continuous basis. The process is in commercial use for particulate Al-based composites (Skibo *et al.* 1988) and the material produced is suitable for further operations such as pressure die-casting (Hoover 1991). Adaptation of conventional melt handling techniques allows the viscosity increases caused by the presence of up to about 25% of particulate to be accommodated. More problematical are sources of microstructural inhomogeneity, including particle agglomeration and sedimentation in the melt and redistribution as a result of particle pushing by an advancing solidification front, which is still poorly understood and can be difficult to eliminate. A typical microstructure illustrating the effect of particle pushing during solidification is shown in Fig. 11.9.

Stir casting usually involves prolonged liquid/ceramic contact, which can cause substantial interfacial reaction. This has been studied in detail (Lloyd 1989) for Al/SiC, in which the formation of Al_4C_3 and Si can be extensive. This both degrades the final properties of the composite and raises the viscosity of the slurry, making subsequent casting difficult. The rate of reaction is reduced, and can become zero, if a Si-rich melt is used. The reaction kinetics and Si levels needed to eliminate it are such that casting of Al/SiC_p involving prolonged melt holding operations is suited to conventional (high Si) casting alloys, but not to most wrought alloys. This problem is one of the reasons for commercial interest in cast Al/ Al_2O_{3p}, although this system has the penalty of slightly higher densities than Al/SiC.

11.2.3 Spray deposition

Spray deposition was developed by Osprey Ltd (Neath, UK) as a method of building up bulk metallic material by directing an atomised stream of droplets onto a substrate. Adaptation to particulate MMC production by injection of ceramic powder into the spray has been extensively explored (Willis 1988), although with limited commercial success. One of the problems is inhomogeneous distributions of the ceramic particles. Ceramic-rich layers approximately normal to the overall growth direction are often seen. Porosity in the as-sprayed state is typically about 5–15% and secondary consolidation processing is needed.

Thermal spraying involves injection of powder or wire into a high-temperature torch, from which molten droplets emerge at high speed. Compared with melt atomisation techniques, deposition rates (usually $\sim 1\,\mathrm{g\,s^{-1}}$) are slower, but particle velocities (~ 200–$700\,\mathrm{m\,s^{-1}}$) are higher.

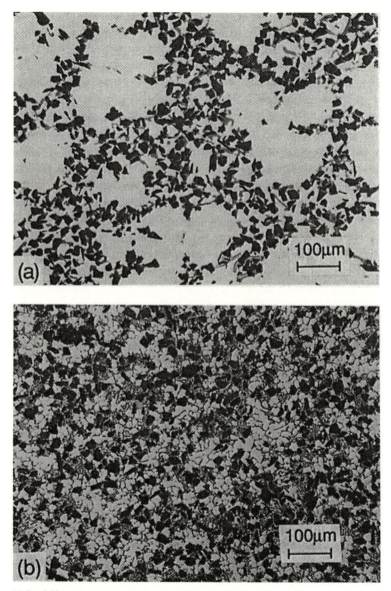

Fig. 11.9   Microstructures (Lloyd 1991) of Al–7 wt% Si/20 vol. % SiC$_p$, (a) investment cast (slow cooling) and (b) pressure die cast (rapid cooling). At the slower cooling rate, SiC particles have been pushed into the interdendritic regions by the growing dendrites, causing severe clustering.

Porosity levels in thermally sprayed deposits are at least a few %. Thermal spraying onto arrays of fibres to form MMCs has received some attention (Clyne and Roberts 1995). An attraction here, particularly for titanium, is that no melt containment is necessary and there is only very brief exposure to high temperatures. Provided the void content and distribution are such that full consolidation could be effected with little further heat treatment, this would allow problems of excessive fibre/ matrix chemical reaction during processing to be avoided. Unfortunately, it has proved very difficult to spray onto fibre arrays so as to produce MMCs with acceptably low void contents (< 10%). Further problems arise from damage to fibre coatings and uneven fibre distributions.

11.2.4 Powder blending and consolidation

Blending of metallic powder and ceramic fibres or particulate has the advantage of close control over the ceramic content. Blending is followed by cold compaction, canning, evacuation, degassing and a high-temperature consolidation stage such as hot isostatic pressing (HIP) or extrusion (Upadhyaya 1989). A feature of much powder route material is the presence of fine oxide particles, usually present in Al-MMCs in the form of plate-like particles a few tens of nanometres thick, constituting about 0.05–0.5vol%, depending on powder history and processing conditions. This fine oxide tends to act as a dispersion strengthening agent and often has a strong influence on the matrix properties, particularly at high temperature. Microstructural changes occur during extrusion. Fibres become aligned parallel to the extrusion axis, but this is often accompanied by progressive fibre fragmentation. The degree of fibre fracture decreases with increasing temperature and decreasing local strain rate. Other microstructural features of extruded MMCs include the formation in some cases of ceramic-enriched 'bands' parallel to the extrusion axis. The mechanism of band formation is still unclear, but it appears to involve the concentration of shear strain in regions where ceramic particles or fibres accumulate.

11.2.5 Diffusion bonding of foils

Titanium reinforced with long fibres is commercially produced by placing arrays of fibres between thin metallic foils, often involving a filament winding operation, followed by hot pressing (Smith and Froes 1984). The procedure is attractive for titanium, since it dissolves its own oxide at temperatures above about 700 °C and the rapid interfacial chemical reaction (Martineau *et al.* 1984) caused by contact between liquid Ti and ceramics is avoided. One of the main problems lies in avoiding excessive chemical reaction at this interface, which tends to lead to embrittlement. There is considerable interest in fibre coatings designed to reduce these problems of interfacial attack.

11.2.6 Physical vapour deposition (PVD)

Several PVD processes have been used in fabrication of MMCs (Everett 1991). An evaporation process used for fabrication of long-fibre reinforced titanium (Ward-Close and Partridge 1990) involves continuous passage of fibre through an evacuated chamber, where condensation takes place so as to produce a relatively thick coating. The vapour is produced by directing a high-power ($\sim 10\,\mathrm{kW}$) electron beam so as to melt the end of a solid bar feedstock. A wide range of alloys can be used; differences in evaporation rate between different solutes become compensated by changes in composition of the molten pool formed on the end of the bar, until a steady state is reached in which the alloy content of the deposit is the same as that of the feedstock. There is little or no mechanical disturbance of the interfacial region during processing. Composite fabrication is completed by assembling the coated fibres into a bundle and consolidating by hot pressing or HIPing. A very uniform distribution of fibres is produced, with fibre contents of up to about 80%. A typical coated fibre and consolidated titanium composite are shown in Fig. 11.10. Production rates are slow and the process is expensive, but it may nevertheless have some commercial potential.

11.3 Ceramic composites

Fabrication of CMCs presents severe technical problems. This is largely due to the brittle nature of ceramic matrices, which makes deformation processing difficult. Furthermore, the matrix is often unable to accommodate the volume changes associated with consolidation without cracks being formed. This is particularly troublesome when fibres are being incorporated, since these tend to resist the contraction of the matrix as voids are eliminated. The need for high temperatures ($> 1000\,^\circ\mathrm{C}$) during most procedures also contributes to the difficulties. There is a limited number of processes which are feasible for production of CMCs, and very few of these are in commercial use. The overview presented in

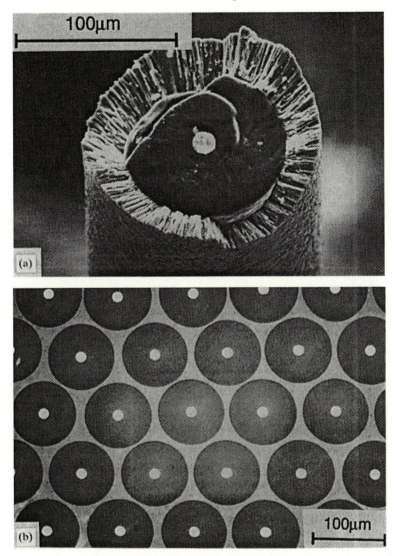

Fig. 11.10   (a) A SiC monofilament with a 35 µm vapour-deposited layer of Ti–5Al–5V and (b) A Ti–5Al–5V/80 vol. % SiC composite produced by HIPing of a bundle of monofilaments with 8 µm thick coatings (Ward-Close and Partridge 1990).

Fig. 11.11 reflects the limited choice available, although production details may vary widely for different materials and applications. Most procedures involve starting material in the form of ceramic powders and the first section below covers general aspects of how these are handled.

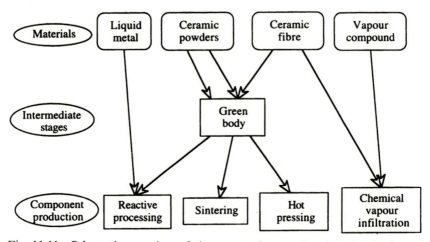

Fig. 11.11 Schematic overview of the approaches employed in fabrication of ceramic matrix composites.

Sections are then devoted to reactive processing methods and to layered composites, neither of which necessarily involve the presence of fibres. Finally, the fabrication of carbon/carbon composites is covered in a separate section, since this case differs from most other CMCs and the process is of commercial significance.

11.3.1 *Powder-based routes*

Production of ceramic artefacts by powder-based routes is very well established in industrial practice. Green bodies are produced by cold compaction of fine powders, often with a binder of some type, and these are then sintered or hot pressed so that voids are eliminated by diffusive processes. In some cases a liquid is present during the consolidation, in which case void elimination is assisted by capillary action and consolidation is much faster. The problems arise when an attempt is made to introduce fibres into such operations. These difficulties are covered in detail by Phillips (1983) and Chawla (1993). Fibres can be mixed with powder particles, for example by blending or by dragging a fibre tow through a suitable slurry. However, subsequent consolidation is then strongly inhibited as the fibres resist the volume contraction of the matrix as it consolidates. Severe matrix cracking, of the type shown in Fig. 11.12, usually results. In some cases, these contraction stresses can be at least partly offset by the imposition of a large hydrostatic

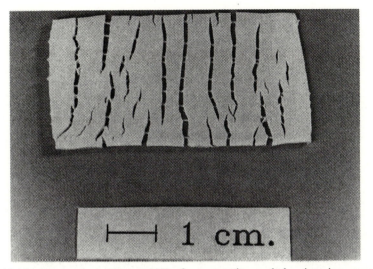

Fig. 11.12   Photograph (Clegg 1992) of a composite made by sintering a powder compact containing dense ZrO_2–3% Y_2O_3 fibres in a matrix of the same material. Large cracks have formed transverse to the fibre direction, along which tensile stresses developed in the matrix as consolidation occurred.

compressive stress, as in the hot isostatic pressing (HIP) process. However, this adds substantially to the processing costs and may not eliminate cracking entirely.

A possible way to resolve this problem is to employ a matrix which is partly, or completely, liquid at the consolidation temperature. This constrains the choice of matrix to those which are likely to show relatively poor properties at elevated temperatures. Nevertheless, there has been interest in making CMCs in this way, using glass or glass–ceramic matrices. Borosilicate glasses and cordierite (glass–ceramic) have attracted particular attention. However, even when the consolidation can be achieved without severe matrix cracking, problems often arise from differential thermal contraction. The misfit strain (see §10.1.1) is often large for systems in which the matrix is fluid at high temperatures, since in such cases it is common for it to exhibit an appreciably higher thermal expansivity than the fibres. An example of the consequences of such differential thermal contraction is shown in Fig. 11.13. In general, the difficulties and constraints on the manufacture of fibre-reinforced ceramics have strongly inhibited commercial development.

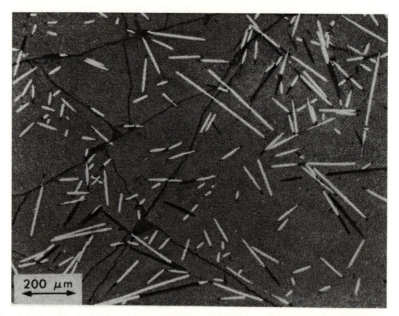

Fig. 11.13   Optical micrograph (Phillips 1983) of a composite made up of short carbon fibres in a matrix of magnesia. Matrix cracking has occurred as a result of differential thermal contraction during cooling after fabrication.

11.3.2 Reactive processing

Several similar processes have been developed in which constituents are brought together under conditions such that a chemical reaction occurs while the mixture consolidates. In several such processes, liquid metal is introduced and progressively oxidises. For example, the directional oxidation of aluminium is exploited in several processes patented under the 'XD' trade name. An attraction of such procedures is that, by making a suitable powder compact into which liquid metal is infiltrated, near-net-shape forming is possible. Levels of internal stresses and porosity can be kept low by control over reaction kinetics, thermal gradients and liquid infiltration rates. In many cases, there is residual unreacted metal, but this can often be tolerated and may help to raise the toughness. Various composite systems have been developed, including several based on intermetallic matrices (Stoloff and Alman 1990). The oxidation reactions exploited by the Lanxide corporation have been covered by Newkirk *et al.* (1987) and an overview of the materials produced by this type of processing has been presented by Urquhart (1991). Some good combinations of properties can be achieved. For example, a SiC-reinforced MoSi$_2$ composite made in this way has been shown (Suzuki *et al.* 1993) to exhibit good creep resistance.

11.3.3 Layered ceramic composites

An attractively simple method of making relatively tough ceramic composites involves stacking thin sheets or tapes in the green state and consolidating these by a sintering operation. The procedures involved have been developed for SiC composites by Clegg and co-workers (Clegg *et al.* 1990, Phillips *et al.* 1994). Fine ceramic powder is blended with viscous polymer solutions and then rolled to produce tapes, typically about 200 μm thick. These tapes are stacked and sintered, after coating with thin ($\sim 5 \mu$m) layers designed to provide weak interfaces which cause crack deflection (see §9.1.2) and hence raise the toughness. Graphitic layers have proved effective, although they suffer from the disadvantage of very poor oxidation resistance at high temperature. The volume contraction on sintering, and subsequent cooling, takes place uniformly and does not result in significant internal stresses. Although the graphitic layers have a different expansivity from the SiC, they have a low modulus and are apparently able to accommodate the shear associated with contraction without becoming damaged. The material is anisotropic, but this is acceptable for many applications (e.g. see §12.8). The microstructure, shown in Fig. 11.14(a), is very similar to that of certain mollusc shells – see Fig. 11.14(b). The production process is relatively quick and cheap, since it does not involve either handling of fibres or application of pressure.

11.3.4 Carbon/carbon composites

Carbon is an excellent high-temperature material, provided it is not exposed to oxidising environments. The excellent mechanical stability of carbon at high temperature, particularly when suitable internal interfaces are present, has led to its use in several important applications (Fitzer 1987), notably aircraft brakes (§12.9). Another consequence of this stability, however, is that it cannot be sintered. There are two basic approaches to production of carbon/carbon composites. Both involve the infiltration of a carbon-bearing fluid into the interstices between an array of carbon fibres. In both cases, the main concern is with achieving complete infiltration in a reasonably short time. Full technical details are given by McAllister and Lachman (1983). The two routes are shown schematically in Fig. 11.15. During liquid impregnation, a pitch or resin

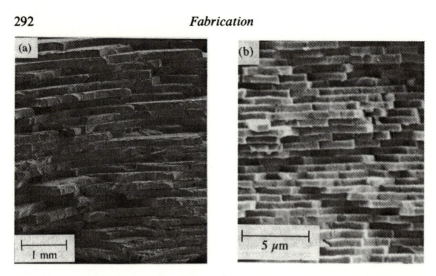

Fig. 11.14   SEM micrographs (Clegg *et al.* 1995) of (a) a layered SiC/graphite composite, with the SiC layers about 200 μm thick and (b) a layered calcium carbonate/protein composite in the mollusc shell pinctada margaritifera.

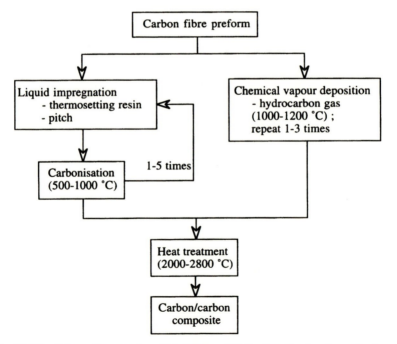

Fig. 11.15   Schematic overview (McAllister 1994) of methods of production of carbon/carbon composites.

is injected and then heated so that it decomposes to leave a carbon deposit. Chemical vapour impregnation (CVI), which can also be applied to various other composite systems (Chawla 1993), involves injection of a suitable hydrocarbon gas, such as methane, together with hydrogen and nitrogen, which decomposes at the infiltration temperature to deposit carbon on the fibres. For both liquid and gaseous impregnation, several cycles of heating and cooling are necessary to complete the operation. Furthermore, it is common to set up a thermal gradient across the component, in order to encourage complete infiltration before the supply of fluid becomes choked off by closing of channels near to the source of fluid. For these reasons, processing is time-consuming and costly.

References and further reading

Becker, W. E. (1979) *Reaction Injection Moulding*. Van Nostrand Reinhold: New York

Blake, J. (1989) *Composite Materials*. Hobsons Publishing: Cambridge

Chawla, K. K. (1993) *Ceramic Matrix Composites*. Chapman & Hall: London

Clegg, W. J. (1992) The fabrication and failure of laminar ceramic composites, *Acta Metall.*, **40** 3085–93

Clegg, W. J., Kendall, K., Alford, N. M., Birchall, D. and Button, T. W. (1990) A simple way to make tough ceramics, *Nature* **347** 455–7

Clyne, T. W. and Mason, J. F. (1987) The squeeze infiltration process for fabrication of metal matrix composites, *Metall. Trans.*, **18A** 1519–30

Clyne, T. W. and Roberts, K. A. (1995) The influence of process parameters on consolidation efficiency when forming titanium composites by spraying onto monofilaments, *Acta Metall. Mater.*, **43** 2541–50

Clyne, T. W. and Withers, P. J. (1993) *An Introduction to Metal Matrix Composites*. Cambridge University Press: Cambridge

Everett, R. K. (1991) Deposition technologies for MMC fabrication, in *Metal Matrix Composites: Processing and Interfaces*. R. K. Everett and R. J. Arsenault (eds.) Academic Press: Boston pp. 103–19

Feest, E. A. (1988) Exploitation of the metal matrix composites concept, *Metals and Materials*, **4** 273–8

Fitzer, E. (1987) The future of carbon–carbon composites, *Carbon*, **25** 163–90

Folkes, M. J. and Russell, D. A. M. (1980) Orientation effects during the flow of short fibre reinforced thermoplastics, *Polymer*, **21** 1252–8

Gotch, T. M. (1994) Spray deposition, in *Handbook of Polymer Fibre Composites*. F. R. Jones (ed.) Longman: Harlow pp. 205–9

Hoover, W. R. (1991) Die casting of Duralcan™ composites, in *Metal Matrix Composites – Processing, Microstructure and Properties*. N. Hansen *et al.* (eds.), Risø National Laboratory: Denmark pp. 387–92

Jones, F. R. (1994) Autoclave moulding, in *Handbook of Polymer Fibre Composites*. F. R. Jones (ed.) Longman: Harlow pp. 138–43

Klier, E. M., Mortensen, A., Cornie, J. A. and Flemings, M. C. (1991) Fabrication of cast particle-reinforced metals via pressure infiltration, *J. Mat. Sci.*, **26** 2519–26

Lloyd, D. J. (1989) The solidification microstructures of particulate reinforced Al/SiC composites, *Comp. Sci. & Tech.*, **35** 159–80

Lloyd, D. J. (1991) Factors influencing the properties of particulate reinforced composites produced by molten metal mixing, in *Metal Matrix Composites – Processing, Microstructure and Properties*. N. Hansen *et al.* (eds.) Risø National Laboratory: Denmark pp. 81–99

Martineau, P., Lahaye, M., Pailler, R., Naslain, R., Couzi, M. and Cruege, F. (1984) SiC filament/titanium matrix composites regarded as model composites. Part 2: Fibre/matrix chemical interactions at high temperatures, *J. Mat. Sci.*, **19** 2749–70

McAllister, L. E. (1994) Carbon–carbon composites, in *Concise Encyclopaedia of Composite Materials*. A. Kelly (ed.) Pergamon: Oxford

McAllister, L. E. and Lachman, W. L. (1983) Multi-directional carbon–carbon composites, in *Fabrication of Composites*. A. Kelly and S. T. Mileiko (eds.) North-Holland: Amsterdam pp. 109–76

Middleton, V. (1994) Filament winding, in *Handbook of Polymer Fibre Composites*. F. R. Jones (ed.) Longman: Harlow pp. 154–60

Mortensen, A. and Cornie, J. A. (1987) On the infiltration of metal matrix composites, *Metall. Trans.*, **18A** 1160–3

Newkirk, M. S., Lesher, H. D., White, D. R., Kennedy, C. R., Urquhart, A. W. and Claar, T. D. (1987) Preparation of Lanxide ceramic composite materials: matrix formation by the directed oxidation of molten metals, *Ceram. Eng. Sci. Proc.* **8** 879–95

Phillipps, A. J., Clegg, W. J. and Clyne, T. W. (1994) The failure of layered ceramics in bending and tension, *Composites* **25** 524–33

Phillips, D. C. (1983) Fibre-reinforced ceramics, in *Fabrication of Composites*. A. Kelly and S. T. Mileiko (eds.) North-Holland: Amsterdam pp. 373–428

Skibo, M., Morris, P. L. and Lloyd, D. J. (1988) Structure and properties of liquid metal processed SiC reinforced aluminium, in *Cast Reinforced Metal Composites*. S. G. Fishman and A. K. Dhingra (eds.) ASM pp. 257–61

Smith, C. S. (1990) *Design of Marine Structures in Composite Materials*. Elsevier: London

Smith, P. R. and Froes, F. H. (1984) Developments in titanium metal matrix composites, *J. Metals*, **36** 19–26

Stoloff, N. S. and Alman, D. E. (1990) Innovative processing techniques for intermetallic matrix composites, *MRS Bulletin*, **15** 47–53

Suzuki, M., Nutt, S. R. and Aiken, R. M. (1993) Creep behaviour of an SiC-reinforced XD $MoSi_2$ composite, *Mat. Sci. & Eng.*, **A162** 73–82

Upadhyaya, G. S. (1989) Powder metallurgy metal matrix composites: an overview, *Met. Mater. Process*, **1** 217–28

Urquhart, A. W. (1991) Novel reinforced ceramics and metals: a review of Lanxide's composite technologies, *Mat. Sci. & Eng.*, **A144** 75–82

Ward-Close, C. M. and Partridge, P. G. (1990) A fibre coating process for advanced metal matrix composites, *J. Mat. Sci.*, **25** 4315–23

Wells, G. M. and McAnulty, K. F. (1987) Computer-aided filament winding using non-geodesic trajectories, in *Proc. ICCM6, Vol.1*. F. L. Matthews *et al.* (eds.) Elsevier: London pp. 161–73

Willis, T. C. (1988) Spray deposition process for metal matrix composite manufacture, *Metals and Materials*, **4** 485–8

12

Applications

Composite materials are used in a very wide range of industrial applications. In this chapter, the objective is to identify some of the considerations involved in commercial exploitation of composites. This is done by means of a few case studies and there is no attempt to present a systematic survey. The examples given cover a range of composite type, engineering complexity, manufacturing route, market size and competitive position relative to conventional materials. At the beginning of each case study, a list is given identifying the reasons for preferring a composite to more conventional engineering materials. Although the examples are spread over the full range of matrix types, the bulk of the annual composite production of around 10 million tonnes is currently in the form of PMCs. At the start of each example, a list is given of the requirements of the application.

12.1 Minesweeper hull

- low density
- ease of moulding to complex shape
- non-magnetic
- good resistance to corrosion and marine fouling
- good resistance to fatigue and stress corrosion cracking

Glass-reinforced plastic (GRP) is now very popular for various land and sea transport applications. While large ships are usually constructed in steel, over 80% of marine hulls less than about 40 m in length are made of GRP (Smith 1990). This is partly because fabrication in GRP is more economic for relatively small craft. Also of importance, however, is the scope for achieving weight reductions and easier maintenance. Additionally, there are certain applications in which the magnetic,

295

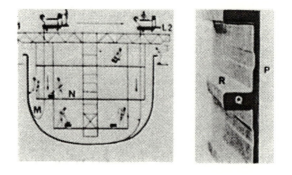

(c)

Fig. 12.1 Automated lay-up of a large marine hull in GRP (Smith 1990). (a) Section through hull, showing gantry for placement and wet-out of glass fibre mats. (b) Stiffening ribs, formed over cores of expanded polymer foam. (c) Schematic perspective view of a minesweeper hull under construction.

electrical or thermal properties of GRP are preferable to those of steel. An example is provided by minesweepers, which need to be non-magnetic in order to avoid activation of magnetic mines.

A hull of this type is usually fabricated by contact moulding, using cold curing polyester resin and E-glass fibre, against an open female steel mould. The mould face is prepared by facing with epoxy, polishing and coating with a release agent. The glass fibre is placed against the mould, commonly in the form of a sequence of plies of chopped strand mat, woven rovings and unidirectional laminae (see §3.2.2). For a fairly large ship, such as a 40-m long minesweeper, the thickness of the lay-up is $\sim$ 30–50 mm. This is too thick for placement and impregnation with resin to be carried out easily by manual methods. A semi-automated arrangement along the lines illustrated in Fig. 12.1 is normally used. Draping of the fibre lay-ups, mixing of resin and catalyst and pumping of the mixture to roller–impregnators is carried out on moveable gantries. Lay-up weights are typically $\sim$ 2 kg m^{-2}; these can be quickly and fully wetted-out using this procedure. Production of a hull with these dimensions is completed in about 10 weeks (Smith 1990).

12.2 Sheet processing rolls

- high beam stiffness ($\propto$ Young's modulus)
- high torsional stiffness ($\propto$ shear modulus)
- low density
- good thermal stability
- good surface finish

High-performance rolls are needed for many processes involving the handling of thin sheets such as newsprint, plastic sheets, etc. In many cases, the rolls need to be relatively wide and so they need a high *beam stiffness* (= moment of inertia of section × Young's modulus of material) in order to avoid excessive bending. A further requirement is that they can be rotated at high speed. This presents difficulties, since rolls with sufficient beam stiffness tend to be so heavy that they become unstable at high rotational speeds. Furthermore, roll speed cannot be changed quickly. In the past, steel rolls were employed, but the above requirements for light, stiff structures have led to their replacment by carbon-fibre/epoxy composites, produced by filament winding (§11.1.1).

A problem with the manufacture of such rolls from fibre composites arises from the need for both bending and torsional stiffness. Torsional stiffness is needed to ensure that the rolls do not distort excessively when a torque is applied to make them rotate. Torsional stiffness is dependent on the shear modulus of the material. The dependence of (axial) Young's modulus, E, and shear modulus, G, on the winding angle, ϕ, for an angle-ply epoxy/65% carbon-fibre composite is shown in Fig. 12.2. A very low winding angle would result in both moduli being low. Peak shear stiffness occurs at $\phi = 45°$, whereas the maximum in Young's modulus is at $\phi = 90°$. At an angle of about 60°, values of E and G are about 70 GPa and 50 GPa respectively, which compare very favourably with steel ($E \sim 210$ GPa, $G \sim 80$ GPa) when account is taken of the fact that the composite has a density which is about one-fifth that of the steel. In industrial usage, weight savings of 75% have been achieved when replacing steel rolls. Rolls up to 9 m long and 0.35 m in diameter have been manufactured. The excellent surface finish required is obtained for the composite rolls by coating them with a thin layer of metal or rubber.

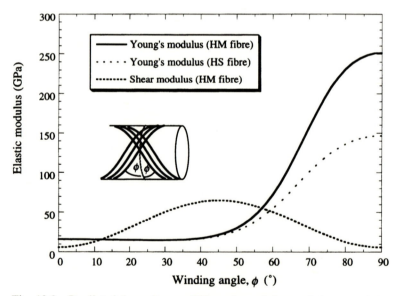

Fig. 12.2   Predicted dependence of Young's modulus and shear modulus on the winding angle, for a tube of angle-ply epoxy/65% carbon (HM) fibre composite, loaded in bending or torsion. A Young's modulus plot is also shown for usage of high-strength (HS) carbon fibres. The plots were obtained by the methods outlined in §5.3.1, using property data given in Tables 2.2 and 2.5. The loading angle for this geometry is given by $(90° - \phi)$.

12.3 Helicopter rotor blade

- high beam stiffness ($\propto$ Young's modulus)
- high torsional stiffness ($\propto$ shear modulus)
- low density
- good fatigue resistance

The rotor blade of a helicopter provides a good example of a component requiring excellent specific stiffness. The blades act as aerofoils which generate lift. A typical rotor blade shape and rotor hub assembly configuration can be seen in Fig. 12.3. Composites have been used for rotor blades, and for other helicopter components, since the 1960s. Initial attractions of using composites included good fatigue resistance as well as specific stiffness. Full use has also been made of the scope for tailoring the elastic properties via control of the fibre architecture and improved aerodynamic blade designs have emerged by stress and fluid

Fig. 12.3   Photograph of a Westland–Agusta EH101 helicopter. (Courtesy of Westland Helicopters).

dynamics modelling, utilising the anisotropic properties of the material (Holt 1994).

A particular problem arises with helicopter blades from the combination of forward and rotational motion. Since the forward velocity of the aircraft may be up to about $100\,\mathrm{m\,s^{-1}}$ and the linear speed of the rotating blade, even at its tip, is often little more than $200\,\mathrm{m\,s^{-1}}$, the airspeed of the blade during the advancing part of the rotational cycle is often substantially greater than that during the retreating phase. If the pitch angle of the blade were the same during each part of the cycle, then the uplift would vary substantially on the two sides of the aircraft and it would be tipped over. Compensation for this effect is achieved by altering the pitch angle of each blade during every rotation. Further changes in pitch angle are used to alter direction during manoeuvring. It is therefore very important that the blades have adequate torsional stiffness, since they must respond quickly and faithfully to pitch-angle changes imposed at the rotor hub. The beam stiffness of the blade must also be high, to ensure that the tip does not lag behind during rotation or flap under its own weight excessively. From this point of view, the requirements are similar to the rolls in §12.2, but the blade presents much greater complexity in terms of sectional shape, loading configuration and fatigue performance. The construction of a typical blade section is shown in Fig. 12.4. The necessary torsional stiffness is provided by the carbon fibres at $\sim \pm 45°$ to the blade axis (see §12.2). The carbon and glass fibres aligned parallel to the blade axis provide the beam stiffness necessary to minimise lag and flap. This construction also confers excellent fatigue resistance.

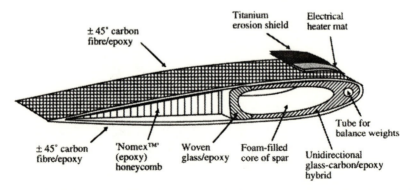

Fig. 12.4 Schematic section through a typical composite construction for a helicopter rotor blade. (Courtesy of Westland Helicopters.)

12.4 Golf driving club

- high beam stiffness ($\propto$ Young's modulus)
- high torsional stiffness ($\propto$ shear modulus)
- low density
- high strength

There are many applications in sports goods with a requirement for stiff, slender beams. These include fishing rods, tennis racquets, skis, surfboards and golf clubs. Polymer-based composites hold a dominant share in all such markets. The sports goods market is one in which small improvements in component performance often justify significant increases in material or manufacturing costs. The golf club (driver) provides a good example of a competitive market with a high premium on performance. As with the previous two case studies, there is a need for the golf club shaft to have good stiffness both in bending and in torsion. High torsional stiffness is very important, since this ensures that the shaft does not rotate significantly under the torque imposed when the club head strikes the ball; any such rotation would introduce an error in the direction of flight of the ball. The requirements for bending stiffness are slightly more complex; some bending of the shaft on impact with the ball may be beneficial, since subsequent straightening can increase the contact time between club head and ball and hence increase the momentum transfer. Designs with a massive club head and slender shaft (see Fig. 12.5(a)) favour increased length of drive. However, in a shaft of low beam stiffness, the axial stresses induced during contact with the ball will be higher and the danger of damage or fracture correspondingly greater. The axial strength of the shaft therefore becomes a key issue.

In view of these requirements, **hybrid** fibre architectures (i.e. involving more than one type of fibre) have been developed. An example of a typical construction is shown in Fig. 12.5(*b*). In this case, the club shaft has several layers of carbon fibre laminae at relatively high angles to the shaft axis. These confer high torsional stiffness. Among the axial laminae are some reinforced with **boron** monofilaments. These have a similar stiffness to carbon (HM), but higher strength (see Table 2.2). In particular, they have a high compressive strength, and are more resistant to buckling than most fibres because of their large diameter ($\sim 100\,\mu m$). These fibres improve the fracture strength on the tensile side of the shaft and, particularly, reduce the danger of failure by kink-band formation on the compressive side. This offsets the disadvantages of higher density and cost of the boron fibres (Buck 1992).

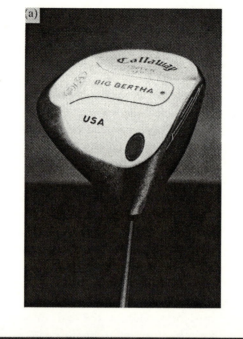

Fig. 12.5 (a) Photograph of the head of a golf club driver. (b) Schematic section through a hybrid carbon fibre/boron monofilament construction for a golf club shaft. (Courtesy of Textron Speciality Materials).

12.5 Racing bicycle

- high stiffness
- good fatigue resistance
- low density

Both DURALCAN and Aerospace Metal Composites (AMC) have developed bicycle frames for commercial sale. The DURALCAN Al(6061)/10%Al$_2$O$_{3p}$ composite material is used in the 'Stumpjumper M2' mountain bike manufactured by Specialized Bicycle Components Inc., while AMC Al(2124)/20%SiC$_p$ material is used for the frame of Raleigh racing bikes. Both models have been successfully tested in extensive sports trials. In the former case, tubing of ~ 1.5-mm wall thickness is extruded and rejoined under pressure around the tube mandrel. The tube sections are then MIG fusion welded, using conventional techniques. In the case of the AMC frame, the material is made by a powder route and then the tubing is adhesively joined. The application of both types of joining procedure to MMCs is described by Ellis *et al.* (1994). In addition to improved specific stiffness, both frames have exceptionally good fatigue endurance, as a result of the enhanced value of ΔK_{th} (see §9.3.1) compared with unreinforced material. Fatigue data for welded tubes have confirmed that the performance of particulate MMCs is superior to that of corresponding unreinforced aluminium (Harrigan 1994). A photograph of the Stumpjumper bicycle in action is shown in Fig. 12.6. Some of the manufacturing procedures involved in producing this bicycle are given by Klimowicz (1994).

12.6 Diesel engine piston

- good wear resistance
- high thermal stability
- high-temperature strength
- good thermal conductivity

This application represents a major early success in the industrial use of MMCs. Production in Japan has been increasing steadily over the past several years and now runs to millions of units annually. Originally, a Ni cast iron (Ni-resist^{TM}) insert was used in the ring area of aluminium pistons in order to prevent seizure of the piston ring with the top ring groove and bore. This impaired heat flow, increased weight and wear rate (Donomoto *et al.* 1983). In 1983, Toyota Motor Co./Art Metal Manufacturing Co. replaced this with a 5% Al$_2$O$_{3f}$ short-fibre insert,

Fig. 12.6 Photograph of the Stumpjumper M2 bicycle in action. The frame is made of Duralcan^{TM} Al(6061)/10% Al$_2$O$_{3p}$. (Courtesy of Duralcan).

thus reducing the weight by 5–10%. This was achieved by squeeze casting (Rohatgi 1991) into an alumina preform to produce a selectively rein-forced component – see Fig. 12.7. In standard tests, wear was reduced by over four times and seizure stress doubled relative to the unreinforced Al alloy. This was combined with four times the thermal conductivity of the Ni-resist^{TM} insert.

Another important factor is the thermal fatigue life (Myers and Chi 1991), which is limited by cracking between the ring groove and the piston itself, or by dimensional instability. The Al$_2$O$_{3f}$ insert outper-forms both the Ni-resist^{TM} insert and the base alloy. The fibre content employed represents a compromise between improved wear and seizure resistance, combined with good machineability relative to the Ni-resist^{TM}, and acceptably small deterioration in fatigue strength and

Fig. 12.7 Photograph (Feest 1988) of a diesel engine piston, showing the region of fibre reinforcement (darker area) in the land (support region) of the groove for the piston ring.

thermal conductivity relative to the unreinforced alloy. Homogeneously reinforced pistons cast from Al/20% SiC_p are also being developed (Rohatgi 1991).

12.7 Microelectronics housing

- thermal expansion matched ($\sim 8 \ 10^{-6} \ K^{-1}$)
- low density
- high thermal conductivity ($> 120 \ W \ m^{-1} \ K^{-1}$)
- electrically conducting

Microelectronic devices often require highly stable environments. For example, microwave radar and communication systems need to be housed so that they are shielded from stray fields and are mechanically, thermally and electrically stable. Kovar^{TM} (high-Ni steel) or brazed steel/molybdenum housings have been used to support ceramic (Al_2O_3) substrate electronic packages. These materials give fairly close matching of thermal expansivity with alumina ($8 \ 10^{-6} \ K^{-1}$). However, they are dense and have low thermal conductivities. This latter point has become a major drawback as progressive increases in component density have led to a greater need for effective heat dissipation.

Possible solutions (Thaw *et al.* 1987, Premkumar *et al.* 1992) include use of Al/SiC$_p$ or Al/boron fibre MMCs. The predicted dependence of thermal conductivity (§10.3.2) and thermal expansivity (§10.1.2) on ceramic content for these two systems is shown in Fig. 12.8. Mapping of these values onto a plot of conductivity against expansivity, as shown in Fig. 12.8(c) for a ceramic volume fraction of 50%, allows identification of materials with good resistance to thermal distortion (high K/α). It can be seen that the SiC particulate reinforcement is most effective. Packaging weight is also reduced significantly ($>65\%$), and machining and brazing distortion minimised ($< \pm50\,\mu m$). In addition, if the housing is cast it is possible to leave an unreinforced region on top of the side walls to aid

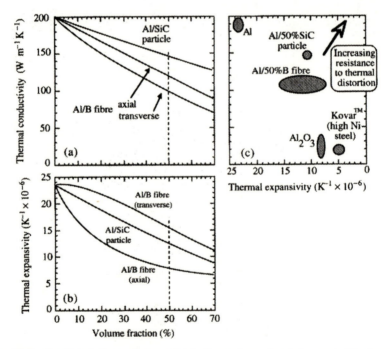

Fig. 12.8 Predicted dependence of (a) thermal conductivity and (b) thermal expansivity on ceramic content for aluminium reinforced with SiC particles or boron monofilaments (long fibres). These plots were obtained by the methods described in §10.1.2 and 10.3.2, respectively, using data in Tables 2.2 and 2.5. Also shown (c) is an Ashby map (§1.2) of the two properties, indicating where selected unreinforced and composite materials are located.

welding of the cover onto the housing. Such components can be cast in particulate MMCs with thin-section walls (Hoover 1991).

12.8 Gas turbine combustor can

- high-temperature strength
- good thermal shock resistance
- low density
- good oxidation resistance
- stable up to $\sim 1450\,°C$

Components at the hot (exhaust) end of a gas turbine are often exposed to very high temperatures and severe thermal shock conditions. The combustor can shown in Fig. 12.9 provides a good example. A fuel/air mixture is fed through the large hole in the end of the can and more air is aspirated through the first set of holes along the sides. The combustion temperatures along the periphery of the flame can reach $1500\,°C$, and the sides of the can rapidly become heated to temperatures close to this level (Schneider *et al.* 1990). At such temperatures, most metals are molten,

Fig. 12.9   Photograph of a combustor can for a gas turbine exhaust system, made of a layered SiC/graphite composite material. (Courtesy of W. J. Clegg)

very soft or, in the case of refractory metals such as molybdenum, prone to rapid oxidation. Some ceramic materials retain good strength and oxidation resistance under these conditions, but are likely to crack under the stresses generated by the high thermal gradients. The can shown in Fig. 12.9 was fabricated from SiC layers ($\sim 200\,\mu$m thick), separated by thin ($\sim 7\,\mu$m) graphitic interlayers (see §11.3.3). The toughness, and the resistance to thermal shock, is raised considerably by the presence of the interfaces, at which any through-thickness cracks become deflected (Phillipps *et al.* 1993).

12.9 Aircraft brakes

- good thermal stability and thermal shock resistance
- low density
- good strength at high temperature
- high thermal capacity
- high thermal conductivity
- good frictional characteristics
- good wear resistance

Aircraft brakes require a particularly demanding set of properties. During an emergency landing or aborted take-off, a very large amount of energy must be absorbed by the brakes in a short time without disintegration or seizure. A typical construction is based on multiple rotating and stationary disks – see Fig. 12.10. Friction between these disks can raise them to average temperatures of 1500 °C, with transient surface temperatures of up to 3000 °C (Chawla 1993). The disk material must therefore have excellent thermal shock resistance and high-temperature strength. Good thermal conductivity is essential to avoid overheating of the disk surfaces. Carbon has good conductivity and high-temperature stability. Solid graphite is a candidate material, which is much cheaper than carbon/carbon composites, but the strength and toughness of the composite is markedly superior. The disks are made using the infiltration techniques outlined in §11.3.4.

The weight of aircraft brakes is also significant. Typically there are eight brakes on a large civil airliner and, when made with a conventional construction (steel against frictional material), these together weigh over 1000 kg. An equivalent set of carbon/carbon brakes weighs less than 700 kg. Weight reductions of several hundred kilograms represent highly significant savings in fuel over the lifetime of the aircraft. Currently,

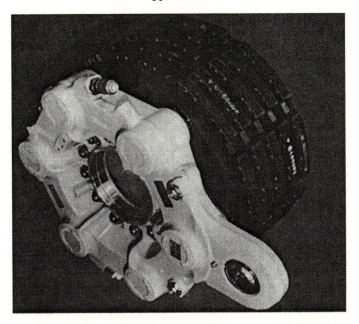

Fig. 12.10   Photograph of the carbon/carbon brake assembly used on the Boeing
767 (Chawla 1993).

carbon/carbon brakes are used in most military aircraft and increasingly
in civil airliners. Concorde is equipped with carbon/carbon brakes, as are
the Boeing 767, 777 and several Airbus models. The major factor inhibiting use of carbon/carbon composites for applications demanding more
prolonged exposure to high temperature is their lack of oxidation resistance, although work is continuing on the development of suitable coatings (Savage 1988).

References and further reading

Buck, M. (1992) Boron fiber: the strength to compete, *Mats. Eng.*, 14–15
Chawla, K. K. (1993) *Ceramic Matrix Composites*. Chapman & Hall: London
Donomoto, T., Funatani, K., Miura, N. and Miyaki, N. (1983) *Ceramic Fibre
    Reinforced Piston for High Performance Diesel Engines*. SAE paper 830252
Ellis, M. B. D., Gittos, M. F. and Threadgill, P. L. (1994) Joining aluminium-
    based metal matrix composites, *Materials World*, 2 415–17
Feest, E. A. (1988) Exploitation of the metal matrix composites concept,
    Metals and Materials, 4 273–8
Harrigan, W. C. (1994) Fatigue testing welded joints for P/M aluminium
    matrix composites, *J. Metals*, 4 (July) 52–3

310 *Applications*

Holt, D. (1994) Helicopter applications of composites, in *Concise Encyclopaedia of Composite Materials*. A. Kelly (ed.) Pergamon: Oxford pp. 125–9

Hoover, W. R. (1991) Die casting of Duralcan™ composites, in *Metal Matrix Composites – Processing, Microstructure and Properties*. N. Hansen *et al.* (eds.) Risø Nat. Lab.: Roskilde, Denmark pp. 387–92

Klimowicz, T. F. (1994) The large scale commercialisation of aluminium matrix composites, *J. Metals*, **42** (November) 49–53

Myers, M. R., and Chi, F. (1991) *Factors Affecting the Fatigue Performance of Metal Matrix Composites for Diesel Pistons*. SAE paper 910833

Phillipps, A. J., Clegg, W. J. and Clyne, T. W. (1993) Fracture behaviour of ceramic laminates in bending, *Acta Metall. Mater.*, **41** 805–25

Premkumar, M. K., Hunt, W. H. and Sawtell, R. R. (1992) Aluminium composite materials for multichip modules, *J. Metals*, **40** (July) 24–8

Rohatgi, P. (1991) Advances in cast MMCs, *Adv. Mats. & Procs.*, **137** 39–44

Savage, G. (1988) Carbon–carbon composite materials, *Metals and Materials*, **4** 544–8

Schneider, G. A., Nickel, K. G. and Petzow, G. (1990) Thermal shock and corrosion of SiC – a combustion chamber model case study, in *The Physics and Chemistry of Carbides, Nitrides and Borides*. R. Freer (ed.) Kluwer Academic Publishers: New York pp. 387–401

Smith, C. S. (1990) *Design of Marine Structures in Composite Materials*. Elsevier: London

Thaw, C., Minet, R., Zemany, J. and Zweben, C. (1987) MMC microwave packaging components, *J. Metals*, **35** 55

Appendix
Nomenclature

Parameters

A	(m^2)	cross-sectional area
a	$(-)$	direction cosine
a	(m)	radius of sphere
a	$(m^2 s^{-1})$	thermal diffusivity
C	(Pa)	stiffness (tensor of 4th rank)
C^{-1}	(Pa^{-1})	compliance (tensor of 4th rank)
c	$(J K^{-1} m^{-3})$	volume specific heat
c	(m)	crack length
d	(m)	fibre or particle diameter
E	(Pa)	Young's modulus
e	$(-)$	relative displacement
f	$(-)$	reinforcement volume fraction
G	$(J m^{-2})$	strain energy release rate
G	(Pa)	shear modulus
h	(m)	spacing between fibres
h	(m)	height
h	$(W m^{-2} K^{-1})$	heat transfer coefficient
I	$(-)$	unit tensor (identity matrix) – see Eqn (6.34)
K	(Pa)	bulk modulus
K	$(MPa\sqrt{m})$	stress intensity factor
K_c	$(MPa \sqrt{m})$	critical stress intensity factor (fracture toughness)
K	$(W m^{-1} K^{-1})$	thermal conductivity
L	(m)	fibre half-length
M	$(m N)$	bending moment
m	$(-)$	Weibull modulus
N	$(mole^{-1})$	Avogadro's number

N	(–)	number of loading cycles
N	(m^{-2})	number of fibres per unit area
n	(–)	dimensionless constant – see Eqn (6.8)
n	(–)	stress exponent
P	(Pa)	pressure
P	(–)	probability
Q	($J\ mole^{-1}$)	activation energy
q	($W\,m^{-2}$)	heat flux
R	(–)	stress ratio
R	(m)	far-field radial distance from fibre axis
r	(m)	radius of fibre or tube
S	(–)	compliance tensor
S	(–)	Eshelby tensor
S	(Pa)	stress amplitude during fatigue
s	(–)	fibre aspect ratio ($2L/d = L/r$)
T	(K)	absolute temperature
T'	($K\,m^{-1}$)	thermal gradient
t	(m)	ply or wall thickness
t	(s)	time
U	(J)	work done during fracture
u	(m)	displacement in x-direction (fibre axis)
V	(m^3)	volume
v	($m\,s^{-1}$)	velocity
W	(kg)	weight
W	($J\,m^{-2}$)	work of adhesion
w	(–)	weight fraction
α	(K^{-1})	thermal expansion coefficient
Δ	(–)	relative change in volume
δ	(m)	crack opening displacement
ϵ	(–)	strain
ϕ	(°)	loading angle (between fibre axis and loading direction)
γ	(–)	shear strain
γ	($J\,m^{-2}$)	surface energy
$\dot{\gamma}$	(s^{-1})	shear strain rate
η	(–)	interaction ratio
κ	(m^{-1})	curvature
λ	(m)	mean free path
θ	(°)	wetting angle

ν	(–)	Poisson's ratio
ρ	(kg m^{-3})	density
ρ	(m)	distance from fibre axis
σ	(Pa)	stress
τ	(Pa)	shear stress
ψ	(°)	phase angle (mode mix)

Subscripts

0	initial
1	x-direction (along fibre axis)
2	y-direction
3	z-direction
A	applied; axis
a	air
b	buckling
c	coated
c	composite
c	critical
d	debonding
e	fibre end
esf	effectively stress-free
f	failure
f	fibre (reinforcement)
fr	frictional sliding
g	global
g	glass transition
H	hoop
H	hydrostatic
i	interfacial
L	liquid
m	matrix
m	melting point
p	pull-out
p	particle
p, q	matrix notation integers
p, q, r	principal stress notations
r	radial
s	survival

RM	Rule of Mixtures
t	stress transfer
th	threshold
trans	transverse
u	failure (ultimate tensile)
u	uncoated
v	volume
x, y, z	Cartesian coordinate directions
Y	yield (0.2% proof stress often taken)
θ	hoop
*	critical (e.g. debonding or fracture)

Superscripts

ax	axial
C	constrained
T	transformation
T*	misfit
tr	transverse

Author index

The work of the author mentioned is cited on the page(s) indicated.

Author index

Subject index

Subject index

Subject index